THE
EVIDENCE BOOK

Comparative Policy Evaluation
Ray C. Rist, Series Editor

THE
EVIDENCE BOOK

CONCEPTS, GENERATION, AND USE OF EVIDENCE
COMPARATIVE POLICY EVALUATION, VOLUME 15

OLAF RIEPER, FRANS L. LEEUW AND TOM LING, EDITORS

Transaction Publishers
New Brunswick (U.S.A.) and London (U.K.)

First paperback printing 2012
Copyright © 2010 by Transaction Publishers, New Brunswick, New Jersey.

This book is printed on acid-free paper that meets the American National Standard for Permanence of Paper for Printed Library Materials.

Library of Congress Catalog Number: 2009022451
ISBN: 978-1-4128-1023-4 (cloth); 978-1-4128-4581-6 (paper)
Printed in the United States of America

Library of Congress Cataloging-in-Publication Data

The evidence book: concepts, generation, and use of evidence / [edited by] Olaf
    Rieper, Frans L. Leeuw, and Tom Ling.
        p. cm. — (Comparative policy evaluation ; v 15)
    Includes bibliographical references and index.
    ISBN 978-1-4128-1023-4 (alk. paper)
        1. Organizational effectiveness—Evaluation. 2. Total quality management in government—Evaluation. 3. Public administration—Evaluation. 4. Quality assurance—Evaluation. 5. Evidence. 6. Comparative government. I. Rieper, Olaf. II. Leeuw, Frans L. III. Ling, Tom.

JF1525.T67E95 2009
352.3'4—dc22

                                                    2009022451

Contents

Preface

Ray Pawson

Let us begin this hugely contemporary book in 1956. More than a half a century ago, the British philosopher Walter Gallie gave a lecture to the Aristotelian Society in which he coined the term "essentially contested concepts." Since that time academics have had a love/hate relationship with the idea and the stormy affair continues into this volume. Gallie had in mind certain concepts upon which there is widespread consensus on their value and utility but no agreement on how to realize or measure them. He was thinking of notions from the political sphere such as "democracy" and "social justice," which he sought to distinguish from the sturdy concepts of science with their clear empirical referents.

The neologism, however, had a life of its own. It became a debating weapon according to which any body of knowledge could be challenged and its key terms declared "polysemic" or "in the eye of the beholder." It appears that post-modernism was born in 1956. In other wiser corners of the academy another reaction prevailed. To be sure there will always be contention and dispute in science and social science, but the point is to examine the nature and source of the disagreements. Knowledge grows as ideas are tested out against each other. Agreement is not rooted in the naming of concepts but in the process of thesis, antithesis, and synthesis.

There are many echoes of these debates in the present volume, though the disputants have spread well beyond the ivory towers. Readers will meet wily politicians and their slippery ideas, but they will also encounter a more surprising and bitter contest on the nature of evidence itself. Where should we locate the "evidence" in "evidence-based policy"? On the one hand there are claims for practitioner wisdom and the voice of experience. Against this is posed the authority of experimental science and the gold-standard of the randomized controlled trial. Battle commences. But this is no stand-off, no impasse. Perhaps there is something to be leaned from common sense and science? Many wise words are uttered in these pages about the potential for such a synthesis.

The other great feature of *The Evidence Book—Concepts, Generation and Use of Evidence* is its breadth of authorship. This is a truly global collaboration. In program evaluation we have come to appreciate the importance of context; success in one location is often hard to replicate in the next. Exactly the same applies the "evidence movement" itself. The willingness to engage in evidence-based policy and the wherewithal to do so is heavily constrained by economic, political, and cultural climates. Evaluation logistics, we learn, need to be tailor-made to national and international capacity.

Rieper, Leeuw, Ling, and colleagues have treated us to a marvellously comprehensive and utterly modern treatise on evidence-based policy. Hopes have been set too high for the evidence movement and this book offers some welcome modesty and realism. The bedrock of evidence is not a body of data or a set of research techniques but a voyage of mutual learning.

I invite readers, as I did, to enjoy the journey.

Ray Pawson
Professor of Social Research Methodology
University of Leeds
UK

1

Introduction

Frans L. Leeuw, Tom Ling, and Olaf Rieper

Discussing Evidence

The contributors to this book are concerned with collecting, ranking, and analyzing evidence and using it to deliver evaluations. As a group, we are aware that the concept of evidence has been used increasingly in the last decade. As with other "magnetic concepts" the concept itself escapes a precise definition. Indeed part of its power is that, while it is seen to be a "good thing," it is sufficiently malleable to appeal to a variety of constituencies. In some instances "evidence" means "empirical data or findings." In other instances it means "information that we can trust." In either case, during the last 20 years many policymakers and practitioners have more often turned to "evidence" to justify and support their actions. Equally "evidence" has been more overtly used in the development of policies, professional guidelines, codes, regulations, and so forth. This growing importance of evidence has been advocated with some enthusiasm by supporters who see it as a way of increasing the effectiveness and quality of decisions and of professional life. This group of advocates lies at the heart of the "evidence-based movement." This is a movement that intends to make evidence more central to decision-making in policy making, health, economics and social well-being and simultaneously identifying what should count as evidence. This movement has touched the work of all the contributors to this book. It has raised important questions for their work as evaluators and academic researchers. These items lie at the heart of this book.

As indicated, there is no single agreed concept of "evidence" Thomas (2004: 1). It is emphasized that evidence takes different forms and is valued differently in different contexts, such as the legal system, natural science, medicine and the humanities. The evaluation community, like any other scientific community, operates with a set of beliefs about what counts as knowledge, what the role of "theories" is, how to collect data, how to weigh different forms of evidence, and how to construct logical and compelling arguments. This shared set of assumptions and beliefs allows evaluators to engage in peer review, argumentation, criticism, and to systematically build a body of shared knowledge. The crucial institutions that help to build this community are journals, evaluation associations, conferences, academic institutions, and evaluation groups inside larger organizations. The processes used are education and training, accreditation, peer review, and critical engagement with each other's work. In the evaluation community, these institutions and processes have become increasingly global in the last 20 years, and this book is a testimony to that. However, the evaluation community differs in an important respect from other scientific communities. The practice of evaluation has for some time been thought of as a meta-discipline (Scriven, 2003; 2005; Coryn and Hattie, 2006) meaning that it draws upon the results of other disciplines in arriving at judgments within its own practice. For example, an evaluation of an intervention to improve the quality of health care will inevitably draw upon the work of health economists, physicians, behavioral psychologists, health sociologists, and patient groups. Therefore, evaluators must concern themselves not only with what they consider to be evidence but also with how *other disciplines* address this question. This book seeks to examine and explain the intersection between the practice and theory of evaluation and the rise of the evidence movement.

We can distinguish between different forms of evidence, e.g. between research-based or research-led evidence and practice-based evidence (Eraut, 2004: 92). In spite of these variations, three criteria are common for judging evidence: (1) relevance in relation to a given assertion, (2) sufficiency, in the sense of corroboration with other instances of the same kind of evidence or other kinds of evidence, and (3) veracity, meaning that the process of gathering evidence has been free from distortion and as far as possible uncontaminated by vested interests (Thomas, 2004: 5). Linked to these is the criterion of internal and external validity. The Maryland Scientific Methods Scale (MSMS), well known in the field of crime and justice evaluations, is an example. This scale describes which types of information from evaluations and other research can be used (or cannot be used) as evidence with regard to the impact of programs and policies on society. The MSMS has five levels, the highest defining the experimental design, and the lowest a correlational design. Underlying this hierarchy of designs is the way in which the attribution problem has been addressed: many believe that in an experimental design this problem is dealt with much more satisfactorily than in a non (quasi) experimental design. This is because the counterfactual behavior ("what would have happened had there been no policy or program") is described by the behavior of the control group (Cook, 2003). This claim has been fiercely discussed and would not be accepted by all evaluators. However, policymakers attach much more value to "experimental evaluations" (see Pawson and Tilley (1994; 1997) for a discussion of his claim and see van der Knaap et al. (2008) for an effort to combine realist evaluation standards and Campbell Collaboration standards). The value and relevance of experimental designs are also debated within the evidence movement. Several chapters of the book present arguments for and against experimental designs and discuss their promises and limitations.

In the area of health evaluations, the so-called evidence-based health sciences (EBHS) in the Anglo-Saxon world have at its core the Cochrane database of systematic reviews. This knowledge repository is the result of the first systematic approach to evidence collection in health care, although it is sometimes still criticized for the suggestion that the database would miss up to 98% the literature on health care as imperfect (Holmes et al., 2006). The Cochrane database not only insists that evidence should underpin decision-making in health care, but also (in line with the Maryland Scientific Methods Scale) privileges knowledge produced through experimental research. However, even if this definition were to be accepted, the consequences of EBHS are harder to predict than many of its critics or its supporters anticipate. In practice, finding the evidence has only established the terms of the debate and not the answers. For example, policymakers saw it as an opportunity to ration scarce resources while the buoyant health consumer movement in the United Kingdom (Allsop et al., 2006) used new evidence as an opportunity to demand the provision of new treatments. In other words, the evidence-based movement did not *end* arguments about what should be done in health care. To expect that would of course also have been too ambitious. However, what can be said is that in fields like crime and justice, labor participation, education and development aid programs, increasingly attention is paid to systematic reviews, randomized controlled trials (RCT), and evaluation synthesis work.

Furthermore, in health care the evidence-based movement has not been given a free rein to define "evidence" in its own image. Even in the United Kingdom, where the evidence-based movement is relatively strong, the

Economic and Social Research Council's Evidence-Based Network was widely resistant to a narrow definition of "evidence" and often argued for the importance of practitioner's knowledge, user knowledge and governance knowledge (Locock and Boaz, 2004). Many remain persuaded that RCT-type evidence is appropriate for addressing some questions, but not all, and that these forms of evidence are not in a hierarchical structure ranging from high quality to low. Rather there is a variety of types of evidence that might be more or less fit for different purposes.

The Systematic Review

Developments in health have been mirrored in areas such as criminal justice, education, social work, and drug control. In these areas, the development of a range of research findings has given rise to one of the truly novel contributions of the EBM—the systematic review. In turn, systematic reviews require a clearly focused definition of evidence. The results can be seen in "systematic reviews" of the kind produced by the Cochrane Centre, the Campbell Collaboration, and the English EPPI (the Evidence for Policy and Practice Information and Coordinating Centre). Distinctions can be made between different types of reviews, such as systematic reviews, narrative reviews, conceptual reviews, realist reviews and critical reviews (for a comprehensive list see Petticrew and Roberts, 2006: 39ff). Most recently, Pawson (2006) published a monograph on how to do synthesis work in a realist tradition. This book is a questioning of the Cochrane type of meta-analysis and advocates in its place a focus on the (behavioral) mechanisms, and the context in which they function. For example, a realist approach to synthesizing evaluations would focus on a policy intervention (for example, naming and shaming of sex offenders or offering financial inducements for energy conservation behavior), then find out which social and behavioral mechanisms are assumed to be at work (damaging reputations, community effects, installing shame in the persons publicly "named" which reduces "recidivism") and consider how these mechanisms had worked in a variety of contexts such as community safely, education, health care and so forth. A Cochrane review would, on the other hand, focus on reviewing interventions in the same policy area (for example, efforts to address a particular health problem).

A systematic review in the Cochrane sense is defined as a summary of the results of existing evaluations and research projects, and it is produced with the purpose of clarifying whether given interventions work in relation to a given social problem. They aim to reduce bias and increase confidence when drawing conclusions. It is clear that only reviews that explicitly account for the criteria which have been used to identify and select studies are to be included in the reviews. However, there is still some conceptual uncertainty related to the concept of systematic reviews. Sometimes it is used narrowly to cover reviews merely including statistical summary (so-called meta-analysis) of the included studies. In line with Petticrew and Roberts (2006: xiv) we use the concept of systematic reviews to cover both reviews including RCTs, statistical analysis, and reviews building on other designs of primary studies and other methods of synthesizing.

In this book we identify several definitions of "evidence" covering meaning in scope from "empirical research findings" to "systematic reviews." Each chapter will pay attention to this definitional problem.

Evaluation and Evidence

Evaluation and evidence are interconnected. First, because evaluations (among other kinds of studies) produce and depend upon evidence (that is, empirical research findings). Furthermore, systematic reviews draw together evidence on the effects of interventions in ways that are helpful for evaluators. *One question* that this book addresses is whether "evidence-based evaluations" is just another term for good quality evaluations, or is it intended to mean something else? It would, after all, be hard to describe an evaluation which did not use or produce evidence. *A second question* is whether evidence in systematic reviews represents a distinct order of knowledge. In this sense, "first order" knowledge comes from evaluations that use natural persons and corporate actors, while "second order" knowledge involves the integrating, synthesizing, weighing, and re-ordering of existing data and analyses before claiming this to be a more reliable source of evidence. Is synthesized or aggregated evidence better than, or merely different from, its constituent parts?

A *third question* is: does the term "evidence-based policy and practice" add anything new to the term "evaluation-informed public policy and practice"? Then, if "evidence" adds any new dimension to evaluation, what is this and why is it "new." Another challenging question is how the utilization of this evidence takes place by decision-makers and citizens? This is the *fourth question* of the book. Has, after the rise of the evidence-based movement, the availability of evidence to decision-makers made decision-making easier and more to the point? Should we regard evidence as something that supports learning or is it a substitute for learning—does the evidence speak for itself or does it need to be located within a complex world of multiple values and uncertain causality?

We also ponder on the methodological issues and controversies behind the evidence movement. Are we witnessing a new "war of scientific paradigms"? Or are we observing another round of the old methodological debate between *verstehende* and *erklärende* sciences with no really new dimensions added, but only disguised in old clothes? The idea of a hierarchy of evidence obviously raises the idea that "hard" quantitative sciences produce "better" evidence than "soft" qualitative sciences. Is this declaring victory to one side of the methodology conflict a sleight-of-hand? Although it would have been possible, we do not discuss underlying *epistemological topics*. One such topic being what is the role of the *covering law model*, a term coined by Hempel and Oppenheim (1948), when talking about evidence. This model stresses deduction and the role of theories ("laws") but allows the use of different types of evidence when testing theories. However, such a topic would have been interesting but unfortunately lies beyond the focus of this book.

The volume visits a number of policy sectors covering health, education, social affairs, development aid and crime and justice. And the book spans several nations of different kinds. The question is whether the nations as well as the sectors provide different contexts for how evidence is defined, constructed and used? For example, why has the evidence-based movement enjoyed a higher profile in Anglo-Saxon countries, and why has the world medicine been more receptive than, say, social care? This is the *fifth question* of the book.

Looking Forward

This leads us to the final, *sixth question*: is the evidence-based movement an empty balloon, impressive from the outside, but unable to sustain its shape when pricked by critical scrutiny? Conversely, is it a dangerous concept that excludes relevant other high quality knowledge. Or will it contribute to improved evaluative information relevant for, and used by, citizens and decision-makers for the common good? And what wider consequences do we foresee?

It is the ambitious objective of the book to address these six sets of questions.

The Content of the Book—In Short

The book is structured in three parts. Part One provides conceptual clarifications and traces the history of the evidence movement and its recent developments. Part Two describes and discusses various forms of evidence construction and generation. Part Three analyses and discusses the application of evidence in policy and practice, and draws together conclusions and prospects.

In Part One *Frans Leeuw* traces the roots of evaluation and evidence and discusses the experimental orientation in social science disciplines such as sociology and psychology and within policy sectors especially in education, crime and justice and social policy. He discusses the ups and downs of experiments in social sciences especially as they relate to policy and program evaluation. He also addresses the question whether or not there is a difference between the type of evidence produced then and now. *Hanne Foss Hansen* and *Olaf Rieper* present an overview of the recent adoption, organizational construction and practice of the systematic reviews in Europe based on the conceptual framework of institutional analysis. Reasons for the rapid increase in evidence-producing organizations are discussed.

In Part Two *Tom Ling* asks what constitutes evidence for decision-making within different policy sectors (such as health, education, development aid and security) with various "epistemic communities." How are evaluation communities able to learn from each other across policy areas? And he discusses whether a general understanding of evidence (shared standards) is emerging across sectors—and whether it is something to strive for. *Jos Vaessen and Rob van den Berg* address the challenges of generating evidence in evaluations of multi-level interventions in development aid. Their example is the Global Environment Facility (GEF), that helps developing countries fund projects and programs that protect the global environment. The end users' perception of effects of interventions in health care is compared and discussed in relation to "independent-of-user evidence" by *Laila Launsø and Olaf Rieper*. Measurement as defined by professionals seems not to catch up with the assumed "wholeness" of the perceptions by end users.

In Part Three, *John Mayne* takes as a point of departure that result-based management supposed to bring evidence to public management decisions has not been successful. He takes a closer look at the kind of evidence used in result-based management and investigates the "theory of result-based management." Finally, he presents a revised, normative theory. This is followed by a chapter by *Richard Boyle* which addresses the paradoxical question why politicians and citizens seem to like using evidence that may not count as evidence. His point of departure is politicians' use of international comparative data on public sector performance and school students' use of the "www.ratemyteacher.com" (or similar sites). He also discusses what prompts politicians and citizens to use "good" evidence.

The volume finishes with the editors providing an outlook for future trends: Is "evidence" a fading mode, or is it a sustainable movement that really does address urgent needs of our knowledge society?

References

Allsop, J., K. Jones and R. Baggott (2006). "Health consumer groups in the UK: a new social movement?" *Sociology of Health and Illness* 26: 737-56.

Cook, T.D. (2003). "Why have educational evaluators chosen not to do randomized experiments?" *Annals of American Academy of Political and Social Science* 589: 114-49.

Coryn, C.L. and J.A. Hattie (2006). "The transdisciplinary model of evaluation." *Journal of MultiDisciplinary Evaluation* 4: 25-40.

Eraut, M. (2004). "Practice-based evidence." In G. Thomas and R. Pring (eds.), *Evidence-Based Practice in Education*, New York: Open University Press: 91-101.

Hempel, C.G. and P. Oppenheim (1948). "Studies in the logic of explanation." *Philosophy of Science* 15: 135-75.

Holmes, D., S.J. Murray, A. Perron, and G. Rail. (2006). "Deconstructing the evidence-based discourse in health sciences: truth, power and fascism." *International Journal of Evidence Based Health* 4: 180-86.

Leeuw, F.L. (2005). "Trends and developments in program evaluation in general and criminal justice programs in particular." *European Journal on Criminal Policy and Research* 11: 18-35.

Locock, L. and A. Boaz. (2004). "Policy and practice—Worlds apart?" *Social Policy and Society* 3: 375-84.

Pawson, R. (2002). "Evidence-based policy: The promise of Realist Synthesis." *Evaluation*, 8: 340-58.

Pawson, R. (2006). *Evidence-based policy. A realist perspective*. London: SAGE Publications Ltd.

Pawson, R. and N. Tilley. (1994). "What works in evaluation research?" *British Journal of Criminology* 34: 291-306.

Pawson, R. and N. Tilley (1997). *Realistic evaluation*. London: Sage.

Petticrew, M. and H. Roberts (2006). *"Systematic reviews in the social sciences: A practical guide."* Blackwell Publishing.

Scriven, M. (2003). "Evaluation in the new millennium: The transdisciplinary vision." In S.I. Donaldson and M. Scriven (eds.), *Evaluating social programs and problems: Visions for the millennium*, Mahwah, NJ: Lawrence Erlbaum Associates: 19-42.

Scriven, M. (2005). *The transdisciplinary model of evaluation: Radical implications*. Paper presented at the meeting of the American Evaluation Association/Canadian Evaluation Society, Toronto, Canada.

Thomas, G. (2004). "Introduction: evidence and practice." In G. Thomas and R. Pring (eds.), *Evidence-Based Practice in Education*, New York: Open University Press.

van der Knaap, L.M., F.L. Leeuw, S. Bogaerts and L. Nijssen (2008). "Combining Campbell standards and the Realist Evaluation Approach. The best of two worlds." *American Journal of Evaluation* 29,1: 48-57.

Part I

Concepts and History of the Evidence Movement

2

On the Contemporary History of Experimental Evaluations and Its Relevance for Policy Making[1]

Frans L. Leeuw

Introduction

If I would have been a historian, the origins of empirical sociology (Comte, Quetelet, the first part of the nineteenth century) and empirical psychology (Wundt, Brentano, second part of the nineteenth) would have been a good starting point for a chapter on the contemporary history of the evidence-based "movement." Comte emphasized a positivist(ic) quantitative, mathematical basis for decision-making, while Quetelet applied—then—new statistical methods in the social sciences. His *Research on the Propensity for Crime at Different Ages* (1831) was one of the first large-scale studies to provide a description of the aggregate relationship between age and crime, resulting in the well-known "age-crime curve." Till now, that is one of the core findings of criminological research.[2]

Psychologist Franz Brentano wrote the first textbook about empirical and experimental psychology in 1874, while his colleague Wilhelm Wundt opened the first laboratory for experimental psychological research at Leipzig University in 1879. Without doubt, these founding fathers of social science research were actively involved in producing (empirical) evidence.

However, also work by philosophers of science like Carl Hempel and Karl Popper, psychologist Donald T. Campbell and economists Ludwig von Mises and Friedrich Hayek, who all worked in the twentieth century, can be mentioned. For example, Popper's plea for piecemeal engineer-

11

ing implied a careful monitoring of the consequences of engineering. Campbell's *Experimenting Society* is a direct spin-off of this thinking. Hempel and Oppenheim (1948) suggested the covering law model, pointing to the importance of theories ("laws") and deductive reasoning, when doing research. For evaluators involved in theory-driven work this line of thinking continues to be important. It was Nobel Prize winner of economics Hayek (1944) who showed that planning of policies and their implementation would lead to unintended and unwanted side effects, while von Mises (1929; 1957) suggested the theory that interventions breed interventions. His argument was that although governmental interventions would in one way or the other (always) be ineffective, nevertheless governments continued to implement them because that is what they think has to be done. The "nothing work-phenomenon" which was referred to during the 1960s and 1970s in the United States is a relatively late (re)discovery of this argument (Rossi, 1987).

However, in this chapter I will not go back that far. Following Oakley (2000) an overview of activities relevant to the world of "evidence" (-based policy making) will be given going back to only those activities during the earlier part of the twentieth century in the fields of education, crime, health and social welfare that we would now label as "evaluations" and that were based on "the experimental method." Later, in the 1960s, in the United States, the focus shifted towards public policy programs that had to be evaluated. The 1960s are referred to in the United States as the Golden Age of Evaluation. The "evidence-based policy" movement as we now know it, started later, in the early 1980s, with Archie Cochrane as an important founding ("organizational") father. Other (intellectual) fathers were Campbell (and Cook) but in the earlier part of the twentieth century social scientists like Edward Thorndike and Stuart Chapin also played an important role as knowledge producers.

This chapter summarizes the history of evidence-based policymaking, focusing on experimental (and quasi-experimental) studies. I do so because nowadays it is a strong tendency in the evidence movement that next to the more general, encyclopedic definition of "evidence" ("the available body of facts or information indicating whether a belief or proposition is true or valid" (Oxford English Dictionary), the evidence concept used if one refers to "evidence-based policy" (or medicine, or policing or management) is strongly focused on experimental results.[3] In this definition the focus is on what will tell us what works and what comes from well-designed, well-assessed, controlled trials.

Evaluation and Experiments during the First Part
of the Twentieth Century

Oakley's article on "Experimentation and social interventions: a forgotten but important history" (1998) starts with the following remark. "The research design of the RCT is primarily associated today with medicine." Often it is believed that experimentation, the "evidence movement" and medicine is one and the same. This includes the belief that "evidence-based policymaking" started with Cochrane, the founding father of the Cochrane Collaboration. Oakley (1998; 2000) however makes clear that in the beginning of the twentieth century, at least in the United States, social experimentation, was already an important characteristic of the social sciences.[4] Her chapter on "experimental sociology" during the first part of the twentieth century describes developments in the world of psychology and sociology that can be seen as roots of the current evidence-based movement. She refers to psychologist Thorndike as a leading figure in educational experimentation at Columbia University in the early decades of the twentieth century. His work had several followers, amongst others in the United Kingdom, where Winch used control groups in his empirical research. Schools were seen as "natural laboratories" (Oakley, 2000: 165). However, even to standards-then, not all the experiments were adequately designed. Oakley shows that the continuation of poorly designed studies till the 1940s led to the desire for improvements in experimentation, but a little later also caused the educational field to "move away from experimental work" (Oakley, 2000: 168). As will be discussed later, this cognitive migration of moving away from experimental studies has also happened in the 1970s in the United States.

Also at Columbia University, sociologist Stuart Chapin was active doing "projected designs," a concept he used to describe that "experiments are observations under controlled condition." Four studies that were done in the 1930s and 1940s were examples of these "projected designs." One evaluation focused on the effects of a rural hygiene program, including water boiling, on families' health behavior inside an Arab village. A second study looked into the effects of rehousing on the living conditions of slum families in Minneapolis, while a third study evaluated the impact on social adjustment on encouraging university students to engage in extracurricular activities. The fourth study concerned juvenile delinquency. It focused on the question what the impact was of integrating "normal" and problem children in a program of "controlled activity," such as participation in classes in art, woodwork, or metal. Although the researcher concluded that the program had made problematic children "significantly less problematic," Oakley (2000: 171) is of the opinion that this conclusion is "dubious." Nowadays, these four studies would be seen as contributions to "evidence-based policymaking."

Another interesting line of work in the United States—that is clearly focused on producing "evidence"—can be found in the tradition of "laboratory sociology" (Oakley, 2000: 174). Her topics of research were—again—"troublesome" behavior of boys, although Oakley also refers to a controlled trial with "problematic" adolescent girls at a vocational high school in New York.

During World War II other experimental evaluations were carried out, some of them referred to as the Hovland experiments. Carl I. Hovland worked at Yale University. During the late 1930s and early 1940s he made major contributions to several areas of human experimental psychology, such as the efficiency of different methods of learning. During World War II he left the university for over three years in order to serve as a senior psychologist in the U.S. War Department. His primary role was to conduct experiments on the effectiveness of training and information programs that were intended to influence the motivation of men in the American armed forces. One of the most widely cited of these experiments on opinion change involved testing the effects of a one-sided versus a two-sided presentation of a controversial issue. The results contradicted contentions that claim that a communication that presents only one side of the issue will generally be more successful than one that mentions the opposing side of the argument. These wartime studies were reported in the book *Experiments on Mass Communication* (1949), written jointly by Hovland, A.A. Lumsdaine, and F. D. Sheffield.

Transfer of "experimentation" in the United States to Europe shows an interesting picture. We focus on three countries, the United Kingdom, Germany (Oakley, 2000), and the Netherlands (Leeuw, 2004). The conclusion is that this transfer was limited.

In the United Kingdom the Webb´s *Methods of social study* (1932) had a chapter on experiments but "they reveal their ignorance of development in American social science declaring that it is practically impossible for sociologist to experiment" (Oakley, 2000: 188). She also cites a study which showed that in the 1930s in the United Kingdom there was no teaching about "laboratory sociology." However, since the establishment of the Research Unit within the British Home Office in 1957, experimental evaluations in the field of crime and justice took place. Examples are effectiveness studies of social work in prisons, the effectiveness of probation programs and the effectiveness of a therapeutic community in a residential school for delinquent children between ten and 17 (Nuttal, 2003: 273 ff). These studies "mark both the beginning and the end of random allocation experiments in Home Office research" (Nuttal, 2003: 274). Problems the evaluators had in doing these studies, the mixed bag of results and the problem that "the experiment[s] might have been able to say what had happened but [it] could not answer how or why" (Nuttal, 2003: 277) were the reasons.

With regard to Germany, Oakley finds that although the origins of German sociological research have been important for American "experimental sociology" and evaluations, later the attention for experiments as producers of evidence drifted off in Germany because, as she argues, scholars "saw it as conflicting with humane moral values." However, when in the aftermath of the *Dritter Methodenstreit* within the German social sciences (which concerned the debate between critical rationalism (Popper, Albert, Topitsch a.o.) and the *Frankfurter Schule* with Habermas, Adorno and Horkheimer as leading figures, empirical, behavioral and even experimental sociology in Germany came a bit more to the foreground than in the earlier times. Books were published on *Verhaltenstheoretische Soziologie* [behavioral sociology] (Opp, 1978) that had their roots in studies by Malewski (1967), Homans (1974) and Burgess and Bushell (1969), while at the same time methodological textbooks were published that included experimental designs (Opp, 1976).

With regard to the Netherlands, a mixed perspective can be described (Leeuw, 2004). Although policy research started in the 1920s, the first "research organizations" were established in the early 1930s and started to grow in the 1950s, experimentation has not been high on the agenda for a long time. More attention was focused on field studies, attitude surveys and "socio-graphy." As I indicated earlier (van Hoesel and Leeuw, 2004), one possible explanation is the need for policy researchers to give important stakeholders a say in their work, their methodology, their recommendations and, sometimes, even their results. This may have to do with the characteristic of "polderization" of the country. In order to take care of the continuous water problems (a large part of the country is built below see level and for years the pressure of continues natural population growth caused social problems), all actors involved had to play "their" role in policy studies, leading to consensus (democracy). When doing experiments, less participation from stakeholders and other partners is possible or (methodologically) "accepted." Experiments make clear what is valid and what not, therefore creating something of an imbalance between on the one hand the need of stakeholders, politicians and bureaucrats to be "listened" to and on the other hand the policy researcher or evaluators' drive for knowledge.

However, in the field of social psychological and organization studies experiments were more often carried out. Nowadays, those studies would have been labeled as research producing evidence-based management (Pfeffer and Sutton, 2006). Experiments done by the Leyden Institute for Preventive Health during the 1950s and 1960s concerned the effectiveness of leadership in public sector organizations and companies, the impact of programs desired to curb absence of workers due to illness and the effectiveness of team work within companies and hospitals. This work was internationally well-known (de Haan, 1994).

Evaluation and Experiments during the Second Part of the Twentieth Century

Three major developments from the second part of the last century need to be mentioned.

The first concerns the "golden age" of (experimental) evaluations in the US. The "golden age" was dominated by the "gold standard" of the randomized controlled trial which became in the 1960s and 1970s the "ruling paradigm for evaluation research," as Rossi and Wright say.

The second development started with Archie Cochrane, a British epidemiologist, who published the book *Effectiveness and efficiency* (1972), in which he introduced the concept of evidence-based medicine. Cochrane's criticism was that medicine had not organized its knowledge in any systematic, reliable and cumulative way; the result was chaotic, individualistic, often ineffective, and sometimes harmful. He encouraged health practitioners to practice evidence-based medicine (Danish, 2006).

The third development started in the United Kingdom when during the second part of the 1990s and partly caused by the attention of the then new Labour Government of Tony Blair favored "evidence-based policymaking" (Solesbury, 2001).

The "Golden Age" of (Experimental) Evaluations in the United States

Policy programs that were evaluated as of the early 1960s dealt with poor families on welfare, employment, school drop out and drugs and crime. U.S President Lyndon Baines Johnson played a crucial role in this endeavor when he in 1965 requested that both the "War on Poverty" and the "Great Society-program" should be evaluated systematically.

Although one could argue about the directness of the relationship, Oakley (2000: 200ff) sees a link between the social scientific approach of the U.S. Army, which after World War II continued with studies led by Stouffer on "The American Soldier" on the one hand and the PPB- (Planning, Programming, and Budgeting) approach in the 1960s on the other hand. It was exported to other U.S. departments and government organizations after the Ministry of Defense had experimented with this system. Measuring performance, comparing and contrasting goals with intermediate results and outcomes became "the" standard to follow if one wanted to know what happened and with which effects at the level of (central) government organizations.

During the same years the "golden age of (experimental) evaluations" blossomed. Partly as a result of the "scientifization" of the thinking about measuring the performance of government organizations, partly as a result of the earlier attention paid to experiments in the Army at the end of World War II and even the experiences with "laboratory sociology," outlined earlier. However, a probably more important drive to do experimental evaluations was that Johnson, who initiated the Great Society (program) and the "War on Poverty" wanted valid information on the societal effectiveness of his programs and interventions. Light (2006)[5] describes "tides of reform" through U.S. policy making between 1945 and 2006 and shows that the "scientific management tide" and the "watchful eye tide" are most characteristic for these years. If there is one (U.S.) President that deserves the label as "founding father" of an evidence-based policymaking movement as seen from the government, it is Lyndon Baines Johnson. In 1964 he established the Office of Economic Opportunity to develop strategies for a government attack on poverty problems but that also had a coordinating and stimulating task in the field of evaluation. Under the Johnson regime (1963-1968) laws were developed and implemented which mandated evaluations (usually set at one percent of the program spending). In 1967 the U.S. General Accounting Office was required by amendments to the Economic Opportunity Act to evaluate anti-poverty programs, which development got even bigger when in 1970 the Legislative Reorganization Act required the evaluation of program results and the development of staff capacity to do the evaluations. Over the decades to follow, the General Accounting Office measured time after time how the evaluation capacity developed within the Executive branch, with which results and with which impact (Rist and Paliokas, 2002). Oakley's point that during those years one could speak about "the evaluation industry" is well taken. One of the most important characteristics was the attention paid to experiments. Campbell's seminal paper on "Reforms as experiments" (1969) can be seen as the epistemological underpinning of all this. Not only did the Federal Government move on into more and bigger evaluations during the sixties (and during the larger part of the seventies), there was also a "high-growth industry of private firms, academic departments and research centers all specializing in evaluation research" (Oakley, 2000: 2003). She cites data that between 1969 and 1972 federal spending on evaluations increased from $17 million to $100 million. It was during these years that the concept of "social (policy) experiment" was born: experimental (sometimes quasi-experimental) studies with a longitudinal perspective and focused on topics like the Negative Income Tax and other income maintenance programs, housing allowances, family planning information campaigns, penal experiments, health (insurance) programs, educational experiments and worker adjustment programs (Oakley, 2000; Udry, 1974). Ross (1970: 91) labeled this period as "a path breaking piece of social science research which promises to provide unique guidance in the rational planning of social and economic policy."

Unfortunately, as history has shown, things worked out differently. Although it was clear that an evaluation culture and structure within American society at the Federal level had been established, there also were a number of downfalls. One was that, largely due to computational problems,[6] the results of most of the social experiments came too late for the executive and the legislative branches of government to act upon. While Johnson wanted to establish a Great Society, Nixon received most of the results of the experiments, but was much less (and according to some: not at all) interested in these programs. The differences between political, administrative, and evaluation time (Weiss, 1998) showed to be too big. Secondly, and this has been elucidated by Rossi and Wright (1984), was the problem that it was extremely difficult to design social programs that produce noticeable effects in any desired direction. Oakley (2000: 233) refers to this as the problem of "zero effects." Thirdly, and from the perspective of this book most challenging, is what Rossi called the "Stainless Steel Law of Evaluation": the stronger the evaluation design, the larger the chance that zero or no substantial effect of social programs will be found. All these problems and Nixon's involvement in the Watergate scandal made that the Golden Age did not last long. Aaron's book on *Politics and the professors: the Great Society in perspective* (1978) described the process in which evaluation, from being a "science" enabling politicians to help realize the "Great Society," moved down to like an "art", with only very restricted opportunities to contribute effectively to policymaking (and implementation). However, as in a sequence of studies by the Urban Institute has been made clear, there still is a market for social experiments in the United States (Greenberg, Linksz and Mandell, 2003; Greenberg and Shroder, 2004). These authors list several hundreds social experiments carried out since the 1960s.

The Cochrane and Campbell Movement[7]

In 1972 Professor Archie Cochrane introduced the concept of evidence-based medicine. His criticism was that medicine had not organized its knowledge in any systematic, reliable and cumulative way. He encouraged health practitioners to practice evidence-based medicine. A few years later, a for the evidence-based policy movement crucial term and activity was coined by Gene V. Glass from the Laboratory of Educational Research at the University of Colorado. In 1975 he described a method whereby results from various effects studies could be synthesized. He used the term meta-analysis to describe the "analysis of analyses" or "the statistical analysis of a large collection of analysis results from individual studies for the purpose of integration the findings." More or less at the same time, the concept "meta-evaluation" came into existence, describing the process whereby researchers evaluate the methodological (and procedural) quality of evaluations.

Several years later, in 1992, the first Cochrane Centre opened its doors in Great Britain. The center was funded by the British National Health Service and was active in collecting experiments in the medical field and in sending the message. In 1993 the Cochrane collaboration was launched as an international collaboration organized by this British Centre. The purpose was "making up-to-date, accurate information about the effects of healthcare readily available worldwide." The Cochrane Collaboration is an international not-for-profit collaboration between researchers and practitioners with an interest in improving health care practice. Also in 1993, Cochrane Centres were opened in Canada, the United States and in Scandinavian countries. It was in the same year that the first steps were taken towards introducing the experiences from experimental studies (also known as "intervention studies") in health care in the fields of education and social work into databases and repositories. However, it should be noted that the respective *Digests* published by the Urban Institute have also been paying attention to categorization of experiments and transferring results to policymakers. The UK Social Science Research Unit at the Institute of Education at the University of London was awarded a grant for the development of a database registration of interventions in the field of social work and in education, drawing on inspiration from the Cochrane Collaboration's efforts.

Slowly, but steadily, the interest in a "look-alike" of the Cochrane society developed. Several experiences played a role. One was a speech that the British educational researcher David Hargreaves in 1996 gave at the annual meeting of the Teacher Training Agency in which he criticized educational research. One of the points he made was that this research failed to contribute to basic theory and insight, which is relevant for practice (Danish, 2006).

A second development concerned the establishment of the What Works movement with regard to prison reforms. Martinson's 1974 article "What works? Questions and answers about prison reform" had a large impact in this world and paved the way to more evaluations and results-based management and policies.

A third and more important activity was that in 1996 the U.S. Congress required the Attorney General to provide a "comprehensive evaluation of the effectiveness" of over $3 billion annually in Department of Justice grants to assist State and local law enforcement and communities in preventing crime. Congress required that the research for the evaluation be "independent in nature," and "employ rigorous and scientifically recognized standards and methodologies." It also called for the evaluation to give special emphasis to "factors that relate to juvenile crime and the effect of these programs on youth violence," including "risk factors in the community, schools, and family environments that contribute to juvenile violence." The Assistant Attorney General for the Office of Justice Programs asked the National Institute of Justice to commission an independent review of the relevant scientific literature, which exceeded at that time more than 500 program impact evaluations. A year later the report "Preventing crime: What works, what doesn't, what's promising" was ready (the Sherman Report). It had a methodological appendix in which criteria for assessment of the methodological quality were articulated. Next to "traditional" social science criteria like adequate sampling and the use of (multi-method) data collection techniques, it also presented a (preliminary) version of what is now known as the MSMS (Maryland Scientific Methods scale). This scale took correlational studies as level 1 (lowest level) and experimental designed studies as level 5 (highest level, commendable). Soon after its publication it was widely discussed in many countries. As Lee (2007) remarked: "When I read that report shortly after it was published, I immediately sensed a sea change in the way democracies would talk about crime prevention. No longer would we focus just on ideology. Evidence would soon take a much larger role in the debate. Within a year, officials of at least 12 different nations would consider the report and its policy implications, from Seoul to Stockholm, from Wellington to London"[8].

Partly based on the Sherman Report, in 1999, at a meeting at the School of Public Policy, University College London, it was decided to found the Campbell Collaboration as a sister organization to the Cochrane Collaboration, but focused on social research, educational research and criminology. This Collaboration is named after the American social scientist Donald T. Campbell, referred to earlier. A year later the Campbell Collaboration is officially founded in Pennsylvania. Again a few years later, the U.S. Department of Education established a "What Works Clearinghouse." The aim was to provide educators, policymakers, researchers, and the public with a central and trusted source of scientific evidence of what works in education. The What Works Clearinghouse is operated as a joint venture between The American Institutes for Research and the Campbell Collaboration. Also in those years the Nordic countries, Belgium and a number of other countries set up national Campbell organizations. As Foss and Rieper (in this volume) show also new organizations focused on an efficient transfer of results (from amongst others experiments) like EPPI (the Evidence for Policy and Practice Information and Co-ordinating Centre) in the United Kingdom were established.

As of 2006 the Cochrane Collaboration has published more than 4,500 systematic reviews and is widely recognized in the field of medicine. In October 2006 the Campbell database covering the fields of social research, educational research and criminology results in a list of 102 systematic reviews, either completed or in production.

The UK Evidence-Based Movement

As President Johnson's Great Society Program indicated, the role of politicians cannot be underestimated in the establishment of evidence-based policy making and (experimental) evaluations. The same is true for the United Kingdom, though in a somewhat different way. Salisbury (2001) summarized its recent history. "[The] concept of "evidence-based policy" has been gaining currency over the last decade. The 1999 *White Paper on Modernising Government* clearly adopted it as part of its philosophy: this Government expects more of policymakers. More new ideas, more willingness to question inherited ways of doing things, better use of evidence and research in policy making and better focus on policies that will deliver long term goals. Public policy has caught up with these trends in the worlds of research and practice, endorsed and amplified them.... [This] enthusiasm ... for evidence-based policy contrasts with the near silence of its 1994 and 1995 predecessors on the Civil Service. What happened between these dates was...the coming into office of New Labour. Its stance was anti-ideological and pragmatic. It had a new agenda and so new knowledge requirements. It was suspicious of many established influences on policy, particularly within the civil service, and anxious to open up policy thinking to outsiders." Salisbury (2001) also shows the explicit links between (experimental) evaluations on the one hand and this movement on the other hand. "Systematic reviews are something newer; methodologically rigorous exercises in assessing the findings of previous cognate research in order to synthesize the results. The Cochrane Collaboration has pioneered this work in the medical field and it is now being extended into social policy by the Campbell Collaboration. Both secondary analysis and systematic reviews depend on an awareness of previous research so skills in searching for existing data or findings are also important." Although a clear link between experiments and the evidence-based movement exist, the UK case differs somewhat. First there is a specific and institutionalized focus on knowledge management, utilization and transfer (eg. the establishment of the EPPI Centre and look-alikes). Secondly, there is a rather clear link with a political ideology: "What matters is what works."

Conclusions

The "evidence movement," as seen from the perspective of "evaluations as experiments," goes back much longer than is often believed. It is not a historical "arbitrary" event, or a well planned and organized "breakthrough" or an example of what (global) intellectual leadership by one or a few "evaluators" can achieve. It is the result of the interplay of *three drives*.

The first is the drive of evaluators and research for producing valid, well-assessed, experimental studies, the second is the politicians´ drive for "hard evidence" and the third is the drive for institutionalization of the production and the dissemination and transfer of knowledge to society.

The history is also characterized by *ups and downs*. An example of this is the short history of the "Golden Age of (Experimental) Evaluations" in the United States. Computational problems, the imbalance between evaluation time and political time, and the problem of "zero effects" played a crucial role in the relative demise of the "wonder" years.

Sometimes the ups and downs have more epistemological roots. In the United Kingdom, experimental evaluations in the field of crime and justice were done in the 1950s, but died a couple of years later, partly because the experimentalist had no answers to *why-questions*. Probably they were very involved in designing, doing and reporting experiments and forgot to think about answers policymakers and program officials want. If an experiment shows that a program or intervention has no or only unintended side-effects, the questions policymakers ask are "how to explain this" and "what to do now." These questions apparently were difficult to answer. Pawson and Tilley (1997) suggest that because experimentalists often do not pay enough attention to "theories" underlying policy programs and interventions, this contributes to these problems. The debate on the role theories play when doing experimental evaluations goes on (Farrington, 1998; Pawson and Tilley, 1998; 1998a; Leeuw, Knaap and Bogaerts, 2008).

Historically the most important conclusions however to draw is that experimental evaluations and "the evidence movement" can, to some extent, be characterized as a *"longue duree"* in the world of policy and program evaluation. Remembering Popper's piecemeal engineering and Campbell's plea to work towards an experimenting society, probably society is better with than without such a "longue duree."

Notes

1. The description of start of experimental (controlled) investigations, partly based on Oakley (2002) and Oakley (1998), on *Social experimentation and public policymaking* by David H. Greenberg, Donna Linksz and Marvin Mandell (2006) and on some of my own archives.
2. See Jennissen and Blom (2007) who report about a falsification of the universality of Quetelet's crime curves in particular for Dutch Caribbean people.
3. Needless to say that, of course, there is (still) a debate going on whether or not this indeed is "common wisdom." See also the introductory chapter. Sometimes this debate is ferocious, as a 2007 paper by Holmes et al on deconstructing the evidence-based discourse in the health sciences shows. The authors are of the opinion that "the evidence-based movement in health sciences constitutes a good example of micro fascism at play in the contemporary scientific arena.
4. For reasons of space, I leave out a discussion on experimentation in economics and political science.
5. "Scientific management for those who prefer tight chains of command and strong presidential leadership" and the watchful eye "through an elegant system of checks and balances." See the data on page 9 in Light´s article in which formal documents are used to underpin which "tide" or philosophy was (most) en vogue during the 1960s and first part of the 1970s.
6. This is based on an informal discussion with David Henry during the aftermath of a session on "evaluation roots" in Europe (EES London conference in 2006), which he might even not remember. We discussed what explains that the golden

evaluation years did not have a very long impact. One of the things we agreed upon was that statistics, and in particular multivariate statistics were not as developed as they are now and that the hard and software available to evaluators to analyze their data, were light-years away from what now is available to researchers.

7.   Based on: Danish School of Education, Aarhus University, A brief history of the evidence movement, in: DPU, Quarterly Newsletter – Knowledge, ideas, inspiration, 7 November 2006 issue, Aarhus, 2006.

8.   From a foreword to a report on what the evidence is of restorative justice by Sherman and Strang, published by the Smith Institute in London in 2007. http://www.esmayfairbairn.org.uk/docs/RJ_exec_summary.pdf

References

Burgess, R. and D. Bushell (1990). *Behavioral sociology; the experimental analysis of social process*. New York: Columbia University Press.

Cochrane, A.L. (1972). *Effectiveness and efficiency. Random reflections on health services*. London: Nuffield Provincial Hospitals Trust.

Danish School of Education, Aarhus University (2006). "A brief history of the evidence movement." DPU, *Quarterly Newsletter—Knowledge, ideas, inspiration*, November 7, 2006 issue, Aarhus.

Farrington, D. (1998). "Evaluating 'Communities that care': realistic scientific considerations." *Evaluation* 4,2: 204-211.

Greenberg, D., D. Linksz and M. Mandell (2003). *Social experimentation and public policymaking*. Washington DC: The Urban Institute Press.

Greenberg, D. and M. Shroder (eds.) (2004). *The Digest of social experiments*. Washington DC: The Urban Institute Press.

Haan, J. de (1994). *Research groups in Dutch Sociology*, Ph.D. Utrecht University. Amsterdam: Thesis Publishers.

Hayek, F.A. (1944). *The Road to Serfdom*. Chicago: University of Chicago Press.

Hoesel, P. van and F.L. Leeuw, "Geschiedenis van het beleidsonderzoek in Nederland." In P. van Hoesel, F.L. Leeuw and J. Mevissen, *Beleidsonderzoek in Nederland*, Assen: Van Gorcum: 1-13.

Holmes, D., S.J. Murray, A. Perron and G.Rail (2006). "Deconstructing the evidence-based discourse in health sciences: truth, power and fascism." *International Journal of Evidence Based Health* 4: 180–186.

Homans, G.C. (1974). *Social behaviour, its elementary forms*. San Francisco: Harcourt.

Jennissen, R. and M. Blom (2007). *Allochtone en autochtone verdachten van verschillende delicttypen nader bekeken*. Den Haag: WODC Cahiers 2007-04.

Light, P.C. (2006). "The tides of reform revisited: Patterns in making government work, 1945-2002." *Public Administration Review* 66: 6-19.

Malewski, A. (1967). *Verhalten und Interaktion*. Tubingen: Mohr Siebeck Verlag.

Mises, L. von. (1929). *Kritik des Interventionismus,* Wissenschaftliche Buchgesellschaft, Darmstadt, Gustav Fischer, Jena, 1929.

Mises, L. von (1957). "Die Wahrheit über den Interventionismus." *Monatsblätter für freiheitliche Wirtschaftspolitik* 3:1957): 599-607.

Nuttal, C. (2003). "The Home Office and random allocation experiments." *Evaluation Review* 27,3: 267–290.

Oakley, A. (1998). "Experimentation and social interventions: a forgotten but important history." *BMJ* 317:1239-1242

Oakley, A. (2000). *Experiments in knowing: gender and method in the social sciences*. Cambridge: Polity Press.

Oakley, A., D. Gough, S. Oliver and J. Thomas (2005). "The politics of evidence and methodology: lessons from the EPPI-Centre." *Evidence & Policy: A Journal of Research, Debate and Practice* 1,1: 5-32.

Opp, K.-D. (1972). *Verhaltenstheoretische Soziologie*. Reinbek: Rowohlt.

Opp, K.-D. (1976). *Methodologie der Sozialwissenschaften*. Reinbek: Rowohlt.

Pawson, R. and N. Tilley (1997). *Realistic evaluation*. London: Sage.

Pawson, R. and N. Tilley (1998). "Caring communities, paradigm polemics, design debates." *Evaluation* 4,1: 73-91.

Pawson, R. and N. Tilley (1998a). "Cook-book methods and disastrous recipes: a rejoinder to Farrington." *Evaluation* 4,2: 211-214.

Pfeffer, J. and R.L. Sutton (2006). *Hard facts, dangerous half-truths and total nonsense; profiting from evidence-based management*. Boston: Harvard Business School Press.

Popper, K.R. (1971). *The Open Society and its enemies*. Princeton, New Jersey: Princeton University Press.

Rist, R.C. and K.L. Paliokas (2002). "The rise and fall (and rise again?) of the evaluation function in the US Government." In J.E. Furubo, R.C. Rist and R. Sandahl (eds), *International Atlas of evaluation,* New Brunswick and London: Transaction Publishers: 225-249.

Rossi, P. (1987). "The iron law of evaluation and other metallic rules." In J. Miller and M. Lewis (eds.), *Research in Social Problems and Public Policy*, 4:3-20, Greenwich, England: JAI Press.

Salisbury, W.B. (2001). *Evidence based policy: Whence it came and where it's going*. London: ESRC UK Centre for Evidence Based Policy and Practice: Working Paper 1.

Titchener, E.B. (1921). "Brentano and Wundt: Empirical and experimental psychology." *American Journal of Psychology* 32: 108-120.

Udry, J.R. (1974). *The media and family planning*. Cambridge, MA: Ballinger Publishing Company.

van der Knaap, L.M., F.L. Leeuw, S. Bogaerts and L. Nijssen (2008). "Combining Campbell standards and the Realist Evaluation Approach. The best of two worlds." *American Journal of Evaluation* 29,1: 48-57.

Weiss, C.H. (1998). *Evaluation. Methods for studying programs and policies*. Upper Saddle River: Prentice Hall.

3

Institutionalization of Second-Order Evidence-Producing Organizations

Hanne Foss Hansen and Olaf Rieper

Introduction and Concepts

Systematic reviews, meta-analysis, meta-evaluation, and related concepts have achieved international attention and dissemination as a new form of evaluation and knowledge production since the early 1990s. There is no commonly agreed use of concepts in the area, but we will use the term "systematic reviews" as the general term for summaries of existing evaluations and works of research which are prepared with the purpose of clarifying, for example, whether interventions work in relation to given social problems, what results a particular form of teaching achieves, or what effects certain forms of activation have in relation to getting people out of unemployment. The knowledge which is produced through the compilation of systematic reviews is often designated as evidence. The purpose of producing evidence is to promote well-informed decisions about policy and practice development.

The preparation of a systematic review takes place in a series of phases: (1) formulation of the problem statement, (2) systematic searching, (3) critical evaluation of individual studies, and (4) summarization of the results from the selected individual studies. When the problem statement is specified and it is made clear which type of intervention one wants knowledge about, one seeks to identify all relevant existing knowledge contributions (often works of research published in international scientific journals). Then the individual contributions' methodological foundation is critically evaluated, and it is determined whether they are sufficiently valid to be included in the systematic review in question.

This chapter will take the international organizations producing systematic reviews as a point of departure and show, how the ideas developed in these organizations have traveled and got translated into national contexts. Focus is especially on the UK and the Nordic countries and delimited to welfare services within health, social work and education Thus, other areas in which systematic reviews are produced, such as criminology, agriculture, development aid and environment are not included in this chapter.

Two problem statements will be analyzed:

1)   How have the ideas of evidence-based policy making, delivery, and systematic reviews been spread and institutionalized?
2)   How does second-order knowledge-producing organizations present and perform their review practice? What criteria are used for including primary studies and how are results summarized and synthesized?

The analytical approach is institutional in that it focuses on how the idea travels, how organizations are constructed to implement the idea and the methodological practices of these organizations. In institutional theory there are two positions related to the traveling of ideas. One position put forward by DiMaggio and Powell (1991) states that the travel of ideas leads to convergence and institutional isomorphism defined as homogenization across organizations and organizational fields. Another position put forward by Czarniawska and Joerges (1996), Sahlin-Anderson (1996) and Radaelli (2005) states that the travel of ideas leads to divergence as ideas are translated, edited, and shaped as they travel from one institutional and political context to another. On the basis of the analysis of institutionalization processes related to second-order evidence-producing organizations we will discuss whether we find processes of convergence or divergence. In addition, we will discuss whether the increasing number of second-order evidence-producing organizations adds new aspects to the methodological debate about evidence and review practice.

Methodologically, the analysis is based on publicly accessible documentary data, including websites for organizations which produce systematic reviews, reports from conferences on the topic, interviews with key persons, primarily in Denmark, and publications written by observers of the development of evidence-based policy and practice development primarily in the United States and the United Kingdom.[1] In addition, the authors of this paper have participated in seminars and conferences on evidence topics in Denmark and abroad.

The analysis is structured in the following way. The following section focuses on how the idea of systematic reviews has traveled and become institutionalized; the next section is an in-depth analysis of review practice; the last section gives the conclusion.

The Institutionalization and Spread of Evidence-Producing Organizations

In recent years the production of systematic reviews has been institutionalized in both international and national second-order knowledge-producing organizations.

International Organizations

There are two main second-order knowledge-producing international organizations: the Cochrane Collaboration working within the health field; and the Campbell Collaboration working within the fields of social welfare, education, and criminology. Both collaborations have been initiated by scientific entrepreneurs using international scientific professional networks as the stepping stones for institutionalizing the production of systematic reviews. Both collaborations subscribe to the idea of producing globally valid knowledge about the effects of interventions, if possible through synthesizing the results of primary studies designed as RCTs and using meta-analysis as the form of syntheses.

The Cochrane Collaboration published more than 4,000 reviews and protocols (which are plans for ongoing work) since its establishment in 1993. The idea had its offspring in the 1970s in England where Archie Cochrane spoke for the point that results of all relevant studies concerning given interventions ought to be summarized in reviews. His arguments became a source of inspiration for a group of researchers which among others included David Sackett at McMaster University in Canada, who in the 1980s coined the concept of evidence-based medicine. In the second half of the 1980s, at University of Oxford, Iain Chalmers (who later became one of the key persons behind the Cochrane Collaboration) took initiatives to work out reviews concerning pregnancy and childbirth. As these initiatives became recognized, he initiated international meetings with the purpose of establishing an international review organization. The initiative was welcomed among others by the British National Health Service, which in 1992 funded the English Cochrane Centre. The initiative was also supported by the Swedish institute for medical technology assessment as well as by EU.

The organizational idea of the Cochrane Collaboration is to combine international, regional and thematic units. At the central, international level a steering committee and several institutions dealing with conflict regulation have been established. At the regional level regional centers have been established and at the thematic level review groups, methodological groups and networks are the main knowledge-producing units. During the 1990s the spread of the Cochrane initiative was quite remarkable. By year 2005 the organization included 12 regional centers covering most of the world and 50 review groups.

The Campbell Collaboration, which was modeled after Cochrane's Collaboration, was established in the late 1990s. Since then it has published a little less than 100 reviews and protocols. Both the organizational as well as the methodological idea have been spread in what can be characterized as a normative isomorphic change process based on professionalization (DiMaggio and Powell, 1991). On request from people being active in the Cochrane collaboration the chairman of the Royal Statistical Society Robert F. Boruch in 1999 arranged a number of planning meetings in order to establish a review organization (Schneider 2002, Chalmers 2003a, 2003b). By publishing a publication with the title "Randomized Experiments for Planning and Evaluation" Robert F. Boruch in 1997 had followed in the footsteps of Donald Campbell who in the 1960s argued in favor of the importance of designing new social interventions as experiments in order to be able to evaluate the effects. Planning meetings were hosted by University College in London and the Swedish Social Agency. After these meetings the official foundation of the Campbell Collaboration took place in Philadelphia year 2000 headed by Robert F. Boruch, University of Pennsylvania, and Haluk Soydan from the Swedish Social Agency (Petrosino et al., 2001).

The Campbell Collaboration is organized with an international steering committee and four coordinating groups: one for crime and justice, one for social welfare, one for education and one for communication and internationalization. In addition a users group has been established recently. This group has to work with the involvement of users to secure the relevance of the review work. One regional center, the Nordic Campbell Centre, was established in 2002. Thus, the idea of regional centers has not been as successful in Campbell as in Cochrane.

In addition to the Cochrane Collaboration there are other international second-order evidence-producing organizations within the health field. The Health Evidence Network (HEN) was launched a few years ago by WHO/Europe with the purpose of providing public health decision-makers with trustworthy source of evidence. Since 1999, the BEME Collaboration has worked on systematic reviews of medical education. Systematic reviews are also produced, used and disseminated by units working with technology assessment within health services. The idea of health technology assessment has another source than the idea of systematic reviews, but preparation of systematic reviews is an important element in health technology assessment. Such organizations have been in existence in several countries since the 1980s. In 1985, the International Society for Technology Assessment in Health Care (ISTAHC) was established as a scientific society with the purpose of disseminating and facilitating technology assessment. In 2003, the organization was renamed Health Technology Assessment International (HTAi). The mission of HTAi's is to support the development, communication, understanding and use of health technology assessment around the world as a means of promoting the introduction of effective innovations and effective use of resources in health care (see: http://www.htai.org/abouthtai/mission.asp). HTAi seems, however, more to be a forum for exchanging experience than really an evidence-producing organization. Evidence-producing ambitions are found in the European network. In 2005, 35 health technology assessment organizations throughout Europe gathered in a European network for Health Technology Assessment (EUnetHTA). The aim here is to coordinate efforts including production, dissemination and transfer of HTA results. The initiative is co-financed by the European Commission and the network members.

The National Level

In many countries second-order knowledge-producing organizations have been established at the national level. At the national level there seems to be much more diversity in approaches to the production of systematic reviews across organizations than is the case at the international level. In Europe England has been a pioneer country in institutionalizing the idea of evidence and production of systematic reviews. Both scientific entrepreneurs and organizations in the voluntary sector working within the social welfare field, e.g. the United Kingdom's largest child care charity Barnado's have played a major role in placing the issue of evidence-based policy and practice on the agenda (Macdonald, 2000). As already mentioned the Cochrane Collaboration was developed with its point of departure in an international research collaboration based at Oxford University. Since the beginning of the 1990s the number of second-order knowledge-producing organizations has grown rapidly. Some of the main organizations are the Evidence for Policy and Practice Information and Co-ordinating Centre (EPPI), established 1993 at the Institute of Education, University of London, in order to produce reviews primarily in the field of education, the Centre for Reviews and Dissemination (CRD) established 1994 to support decision-making in the health sector, Research in Practice (RiP) established 1996 to support evidence-informed practice with children and families and Research in Practice for Adults (ripfa) established 2005 to support evidence-informed practice and policy for adults.

Also Blair's Labour Government has been a central player in relation to spreading the concepts of evidence-based policy making and delivery and systematic reviews across policy areas. In 1999 the Blair Government published a White Paper with the title "Modernising Government." In the paper a policy is formulated which emphasizes the desire for political decisions ("policy making") to be based on knowledge about which means and types of efforts work. Learning should be promoted via increased application of evidence and research and via development of new tools for evaluation (Cabinet Office, 1999: 17 and 20). Two years later the government expanded the perspective to include the desire for practice and task management ("policy delivery") to be based on evidence (Cabinet Office Performance and Innovation Unit, 2001; see also the Magenta book: http://www.policyhub.gov.uk/evalpolicy/magenta/guidance-notes.asp).

The political proposals were followed up in part by giving economic support to already existing evidence-producing organizations (e.g. the Evidence for Policy and Practice Information and Co-ordinating Centre (EPPI) at Institute of Education, University of London) and in part by establishing new organizations, among others Social Care Institute for Excellence (SCIE) in 2001.

It is interesting that the English second-order knowledge-producing organizations within the fields of education and social welfare have a broader approach to the concept of evidence than the Campbell Collaboration. Their ideal is not to focus primarily on results from primary studies designed as RCTs and they have developed and used other forms of synthesis than meta-analysis. They have a more pluralistic typology-based approach aiming at including all relevant results from primary studies of good quality.

Four years ago William Solesbury, who has followed the development of the evidence movement in Great Britain, characterized the development as "a peculiarly British affair" (Solesbury, 2001: 6). Without doubt the characterization was correct at that time. However, in 2007 the situation is different. The desire to base policy and practice development on evidence and systematic reviews has spread to a number of other countries.

The idea of evidence and second-order knowledge-producing organizations was rapidly spread to the Nordic Countries. Shortly after the collaborations had been established both the Nordic Cochrane Centre, being a center for the Nordic countries as well as for a number of East European countries, and the Nordic Campbell Centre were located in Copenhagen. In the health field Denmark has also a unit for health technology assessment located at the National Board of Health and a unit producing evidence in relation to the pharmaceutical area.

The Nordic Campbell Centre being located at the Danish National Institute of Social Research has developed a profile primarily concerned with social welfare and less with education and crime. As the Campbell idea traveled into Denmark it was thus translated into the local institutional context de-coupling review practice within the field of social welfare from the fields of education and criminology. In line with this the Nordic Campbell Centre collaborates with organizations in the other Nordic countries also producing systematic reviews in the field of social welfare. In this context the Swedes were the Nordic pioneers when they in 1993 in the Swedish Social Agency established the Centre for Evaluation of Social Service (CUS) later renamed the Institute for Evidence-Based Social Work Practice (IMS). Later in 1997 FinSoc was established in Finland. In Norway Kunnskapssenteret established 2004 is an important link to Campbell. It is interesting that Kunnskapssenteret which is the national knowledge center for the health sector is the Norwegian liaison into the international networks of Campbell, Cochrane as well as organizations working with health technology assessment.

The idea of evidence-based policy and practice within the educational field in Denmark has traveled another way leading to the establishment of a second-order knowledge-producing organization in the field of education in 2006 at the Danish University of Education (now merged into the University of Aarhus as the School of Education). In this case the idea has traveled through the OECD which has played a central role in relation to spreading the ideas of evidence and systematic reviews in education. Centre for Educational Research and Innovation (CERI) has since the mid-1990s (among other activities) conducted reports on the countries' educational research systems and their relations to practice and policy (OECD, 1995). Also CERI has held a number of conferences to improve education research and its links to practice. Conferences were held in Washington 2004, Stockholm 2005, the Hague 2005 and London 2006. Especially the conference held in Washington arranged in co-operation with the American Department of Education's Institute of Education Sciences (IES) and the American Coalition for Evidence-Based Policy had the purpose of examining the possibilities for increasing the effectiveness of education in OECD countries with the help of evidence-based knowledge (OECD, IES and Coalition for Evidence-Based Policy, 2004).

At the Washington conference representatives from the Danish Ministry of Education, the Danish Ministry of Science, Technology and Innovation, and the Danish University of Education learned about the American What Works Clearinghouse (WWC) established in 2002 to provide educators, policymakers, researchers and the public with a central and trustworthy source of scientific evidence of what works in education (http://www.whatworks.ed.gov/whoweare/overview.html). WWC is administered by the U.S. Department of Education's Institute of Education Sciences through a contract to a joint venture of the American Institutes for Research and the Campbell Collaboration. The tight relations between WWC and the Campbell Collaboration also show the fact that Robert F. Boruch is principal investigator in WWC. Not surprisingly, WWC as well as Campbell subscribe to the evidence hierarchy. Surprisingly, however, WWC has developed a national profile only synthesizing results of primary studies carried through in the U.S.

At the Washington conference the Danish representatives adopted the evidence idea and in spring 2006 it was decided to establish a Clearing House at the Danish University of Education. Here we see an example of a coercive isomorphic change process in which both the OECD and two ministries play important roles. In Denmark, where there are no traditions for using RCTs in the field of education the methodology of the clearinghouse immediately became a topic for discussion. The Danish clearinghouse has not yet published any reviews, but it has produced a review policy departing from a typology thinking of evidence and stating that it is planning to work with several methods for synthesizing among others meta-analysis, narrative synthesizes and combined synthesizes. As the idea of review practice has traveled it has clearly been translated to the national context.

Summing Up

The idea of establishing evidence-producing organizations and carrying out systematic reviews for some years has traveled successfully across organizational fields as well as across nation states. At the international level evidence-producing organizations were first established in the health field. Later the idea was spread to the fields of social welfare, education, and criminology. In both cases scientific entrepreneurs were important developers of the idea. Also at the national level the idea has been implemented. Here a range of other actors among others political and bureaucratic regulatory actors have played a role. The dynamics between the international level and the national level takes several directions. The analysis has shown examples both of processes initiated internationally and moving on to national levels and of national actors gathering in international networks to coordinate and strengthen efforts. In the educational field the OECD has acted as an intervening actor speeding up the traveling of the idea across regions and countries. It is evident that the idea has been translated and shaped as it has traveled. Especially at the methodological dimension diversity has increased. This will be further elaborated in the analysis of review practice in the next section.

In the beginning, the forces behind the development of second-order evidence-producing organizations seemed to have been supply driven. In health the professional networks were very active in organizing and providing funding. In social work and education the development process has been more demand driven by ministries and government funding. However, in social work the professional networks have been active in establishing e.g. the Nordic Campbell Collaboration, consequently the supply side was a driving force as well. Overall the driving forces behind the evidence movement seem to have been a combination of supply and demand, depending on the policy area.

Review Practice

Description of Guidelines of Evidence-Producing Organizations

Across the three areas (health, social and education) and the working methods of the various evidence-producing organizations analyses show both similarities and differences (Hansen and Rieper, in print). A reading of guidelines and handbooks from the websites of a number of evidence-producing organizations (see outline Figure 1) shows that most of the organizations have similar guidelines for the work flow in preparing systematic reviews. The work flow is specified in several phases from formulating the review question, preparing the protocol and getting it approved, searching primary studies and assessing their quality, synthesizing results and writing down the review and getting it approved and published. Also the organizations subscribe to common values concerning the use of explicit and transparent methods and replicable and updateable review practices.

While there seems to be a general consensus concerning the working process as such, there are variations and disagreements concerning the content of individual phases of the process, especially the assessment of primary studies for inclusion or exclusion in the review (see Figure 1). The two international organizations Cochrane (health) and Campbell (social and education) advocate for the use of the hierarchy of evidence with the RCT design at the top as the golden standard. The same is the case for the National Institute for Health and Clinical Excellence (NICE), UK. They only accept quasi experimental designs and longitudinal studies as the second best designs. WWC is on the same line. The WWC is administered by the U.S. Department of Education's Institute of Education Sciences through a contract to a joint venture of the American Institutes for Research and the Campbell Collaboration. Thus, the evidence-producing organizations that are associated with the Cochrane Collaboration and the Campbell Collaboration base their assessment of quality of primary studies primarily on the hierarchy of evidence. Others especially the UK organizations SCIE and EPPI accept a broader variety of designs that are not ranked according to a hierarchy of evidence.

Also we find different approaches in relation to how to synthesize. Not surprisingly those who advocate for RCTs also advocate for the use of meta-analysis while those accepting several types of designs are more tolerant as to the use of other types of synthesis. Cochrane, CRD, Campbell, EPPI, NICE and WWC all have guidelines or similar resources for assessing the quality of the specific primary study. Some of the organizations, however, only cover experiment designs, whereas CRD, Campbell, EPPI, NICE, and WWC also bring guidelines for assessing other types of design (see Figure 1). The criteria for assessing experiment designs are far the most developed ones.

Methods Actually Used

The actual methods for synthesizing the results from primary studies are one aspect of review practice that has been less studied. To our knowledge there exists no general overview of the approaches for synthesizing that have been applied in practice. However, within evidence-based medicine the statistical meta-analysis has been the preferred approach to syntheses (Boaz et al., 2006: 481). Especially in England, in recent years there has been some effort to develop and make explicit a broad range of approaches for synthesizing (Dixon-Woods et al., 2004; Boaz et al., 2006). The basic message of these developments is, first, that a variety of approaches exists. Examples of various approaches are illustrated

Figure 1
Outline of Guidelines and Handbooks from Evidence-producing Organizations with Reference to the Basis for the Assessments of Primary Studies. Accessed at the www January 2007

Organization	Title of guidelines Date of publishing Number of pages	Based on a hierarchy of evidence	Criteria for assessment of the specific design	Comments
The Cochrane Collaboration	Cochrane Handbook for Systematic Reviews of Interventions ver. 4.2.6[1] August 2006, 257 pp	Yes, RCT has highest priority. Other types of design are reconsidered to be supportive.	Reference to literature on assessing the quality of RCT	About assessment of RCTs: Reference to Moher et al. 1995.
The Nordic Cochrane Centre	–	–	Refer to the Cochrane –	Collaboration
Centre for Reviews and Dissemination (CRD)	Undertaking systematic reviews of research on effectiveness[2], March 2001, 164 pp	Yes, RCT has highest priority. Other types of design are considered to be supportive. and economic evaluations	Criteria for assessing effectivity studies, accuracy studies, qualitative studies	Discussion on external validity and assessment scales – both numeric and non-numeric scales.
The Campbell Collaboration (C2)	Research Design Policy Brief[3], November 2004, 25 pp design are considered to be supportive.	Discussion on the hierarchy of evidence. Yet, RCT has highest priority. Other types of	Discussion on assessment of quality of individual design	

Figure 1 (cont.)

	Guidelines for the preparation of review protocols ver. 1.0[4], January 2001, 9 pp	RCT has highest rank order. Contributions of qualitative studies are made explicit	No	
	Steps in proposing, preparing, submitting and editing of Campbell Collaboration Systematic Reviews[5], 4 pp	No	No	Flowchart for development of new systematic reviews[6].
Nordisk Campbell Center (NC2)	How to make a Campbell Collaboration Review: Guide 3 how to make a Review, 135 pp	Yes, RCT has highest priority. Other types of design are considered to be supportive.	–	Refer to C2.
The Evidence for Policy and Practice Information and Coordinating Centre (EPPI)	EPPI-Centre methods for conducting systematic reviews ver. 1.0[7] September 2006, 18 pp	No	Yes	A general guideline which allows multiple methodological approaches.
	Book: Using research for effective health promotion, August 2001, pp 176	–	–	Different contributions of knowledge from research and knowledge from practice are discussed.

Figure 1 (cont.)

Organization	Document			
	EPPI-Centre review companion	–	–	Not accessible on the www.
	Guidelines for the REPORting of primary empirical research studies in education (The REPOSE Guidelines)8, 2004, 12 pp	No	Yes	How to report empirical research within education.
	EPPI-Centre Keywording strategy for classifying educational research ver. 0.9.7[9] 18 pp	No	No	Guidelines for keywords for classifying research studies.
Social Care Institute for Excellence (SCIE)	The conduct of systematic research reviews for SCIE knowledge reviews. December 2006 85 pp	No	– ... the review.	Multiple methodological approaches. Criteria of quality appraisal of included studies. Reporting of study quality in the review.
Health Evidence Network (HEN)	–	–	–	HEN has no guidelines.
National Institute for Health and Clinical Excellence (NICE)	The guidelines manual 2006[10] 2006 190 pp	EBM-hierarchy	Assessment form for each study design	Internal and external validity and forms of bias.

Figure 1 (cont.)

IMS – Institute for Evidence-Based Social Work Practice, Sweden (IMS)	–	–	IMS has a guideline in progress.
What Works Clearinghouse (WWC)	Evidence standards for reviewing studies[11] September 2006 12 pp	Yes	Yes

1 http://www.cochrane.org/resources/handbook/Handbook4.2.6Sep2006.pdf
2 http://www.york.ac.uk/inst/crd/report4.htm
3 http://www.campbellcollaboration.org/MG/ResDesPolicyBrief.pdf
4 http://www.campbellcollaboration.org/c2_protocol_guidelines%20doc.pdf
5 http://www.campbellcollaboration.org/C2EditingProcess%20doc.pdf
6 http://www.campbellcollaboration.org/guide.flow.pdf
7 http://eppi.ioe.ac.uk/cms/Portals/0/PDF%20reviews%20and%20summaries/EPPI-Centre_Review_Methods_1.pdf
8 http://www.multilingual-matters.net/erie/018/0201/erie0180201.pdf
9 http://eppi.ioe.ac.uk/EPPIWebContent/downloads/EPPI_Keyword_strategy_0.9.7.pdf
10 http://www.nice.org.uk/page.aspx?o=308639
11 http://www.whatworks.ed.gov/reviewprocess/study_standards_final.pdf

and discussed in Boaz et al. (2006) (narrative analysis, mixed method synthesis, thematic synthesis, realist synthesis, Bayesian synthesis). Second, it is argued that the choice of approach should take a number of factors into account, such as: the research question (of course) but also, for example, the number of primary studies available, the heterogeneity of interventions and the heterogeneity of data. In short, less mechanical choice of approach and more reflexivity on behalf of both the commissioner and the reviewer are strongly recommended.

An analysis have been made of a selection of completed reviews from Cochrane Library (all reviews from 2005: 497 reviews), all reviews from Campbell (C2-RIPE database: 20 reviews), and all reviews from EPPI (74 reviews from 1996-2006). As for the actual inclusion of RCT studies in the reviews the analysis shows that 50% of Campbell's reviews are based solely on RCT studies, 69% of Cochrane's reviews are based solely on RCT, whereas none of EPPI's reviews are based solely on RCT design.

As for methods of synthesizing, the majority of Campbell's and Cochrane's reviews are done by meta-analysis (60% and 68%, respectively). EPPI has most of its reviews done by narrative methodology (85%). Thus, it seems that the evidence organizations mentioned here actually do as they preach.

The Methodological Debate on the Role of RCT in Systematic Reviews

Differences in review practice first and foremost reflect scientific and methodological controversies. In the United States, where systematic reviews have been in progress for more than 25 years (Shadish, Chacón-Moscoco and Sánches-Meca, 2005), a considerable and, at times, quite fervent debate has taken place especially about the evidence concept and RCTs (House, 2004). The methodological discussions are also ongoing in Europe both in medical science (Launsø and Gannik, 2000) and in the social sciences (see for example the controversy between David P. Farrington from Campbell and Ray Pawson, in Farrington (2003) and Pawson (2006)).

There is general agreement about the need for other designs than RCT when the research questions concern process and implementation of an intervention. There is also general agreement about the need for assessing each research design in its own right: RCTs might be conducted weakly or strongly (in fact adherents of RCT have a well known list of threats to internal validity), and so might surveys, case study designs and other designs. The dispute concerns the rationale behind setting up a hierarchy, ranking designs and placing RCTs at the top when the question is about the effects of an intervention.

The main argument for RCT as the gold standard is well known and can be summarized as follows: by dividing the population into two groups (the experiment group and the control group) by randomization all other causal factors than the intervention is "neutralized" (held constant). Both groups' score on the effect variable should be measured before and after the intervention. This clear logic, it is claimed, is not behind any other research design, therefore other designs for measuring effects are considered to be weaker. Commonly raised objections to the use of experimental design refer to the complex nature of social interventions, the technical issues, practical problems and ethical dilemmas associated with random assignment procedures, and the methodological appropriateness of experimental designs (Clarke, 2006: 566ff).

Alternatives to the Hierarchy of Evidence

Typologies or matrices of evidence have been proposed as an alternative to the hierarchy of evidence to indicate schematically the relative contributions that different kinds of methods can make to different kinds of research questions. An example of such typology is shown in Figure 2 (adopted from Petticrew and Roberts (2003) and (2006)).

This typology was originally suggested by Gray (1997) to help health care decision makers determine the appropriateness of different research methods for evaluating different outcomes. However, it might also have a wider applicability outside the health area.

We propose a more generalized table with other categories in the first column (see Figure 3).

Depending on the purpose of the review the most appropriate research designs listed in Figure 3 might be chosen. For stop-go decisions of interventions RCT and longitudinal designs might be most appropriate because of the strengths of these designs to cope with the contra factual. Case studies and field studies are more relevant if the purpose is to enlighten and adjust implementation. Simple and well defined technical interventions (such as medical drugs) are more appropriate for RCT designs than complex interventions which are difficult to delimit from other interventions.

The choice of designs might also depend on the degree of differentiation between the context and the intervention. If the context is supposed to be stable and does not significantly influence the possible effects of the intervention, then RCT is a relevant design. If, however, the context is

Figure 2
An Example of a Typology of Evidence

Design: Research question:	Qualitative research	Survey	Case control studies	Cohort studies	RCTs	Quasi-experimental studies
Effectiveness Does this work? Does doing this work better than doing that?				+	++	+
Process of service delivery How does it work?	++	+				
Salience Does it matter?	++	++				
Safety Will it do more good than harm?	+		+	+	++	+
Acceptability Will children/ parents be willing to or want to take up the service offered?	++	+			+	+
Cost effectiveness Is it worth buying this service?					++	
Appropriateness Is this the right service for these children?	++	++				
Satisfaction with the service Are users, providers and other stakeholders satisfied with the service?	++	++	+	+		

Figure 3
Another Typology of Evidence: Under What Conditions is
Which Design to be Selected?

	RCT	Longitudinal studies	Case studies	Ethnographic field studies
The purpose of the review	Intervention stop/go	Intervention top/go. Adjustment	Understanding implementation. Adjustment.	The actors' meaning and understanding. Adjustment
The type of intervention	Well defined "technical"	Well defined	Complex	Varying
The type of context	Low differentiated	Moderately differentiated	Highly differentiated	Varying
Pre-existing knowledge on cause-effect statistically	Weak	To be modeled	Weak	Weak

supposed to be highly differentiated and hard to delimit from the intervention and its supposed effects, then the case study or the anthropological field study might be a relevant design.

Pre-existing knowledge of the cause-effect mechanisms might be another factor conditioning which design to choose. If that knowledge is rather well developed, then a longitudinal study might be a relevant design because an advanced analysis presupposes statistical modeling.

Conclusion

In this chapter we have focused on how the idea of systematic review has traveled, how second-order evidence-producing organizations have become institutionalized and on how they conduct review practice. The analysis above has shown that the idea of establishing second-order evidence-producing organizations in order to work out systematic reviews is not one homogenous idea, but rather an idea embracing several ideational dimensions. We may distinguish between three dimensions: (1) the basic ideational dimension of summarizing and synthesizing results from primary studies by producing systematic reviews; (2) the organizational ideational dimension of how to organize such work and (3) the methodological ideational dimension of how to carry through review practice.

The analysis has shown that at the basic ideational dimension of summarizing and synthesizing empirical findings from several primary studies convergence has arisen as the idea has traveled. The basic ideational dimension is common across second-order evidence-producing organizations at different levels and in different organizational fields. Convergence at this dimension can be explained by the fact that the idea has common characteristics with ideas which have been found to travel with great success (Røvik, 1998). The idea has been produced by scientific institutions and been socially authorized by organizations and nations considered to be successful. Furthermore, the idea is linked to legitimate goals such as promoting well-informed decisions and ensuring both societal effectiveness and the best available interventions to citizens. In this perspective it is not surprising that the idea of summarizing has become widely recognized.

In relation to the organizational ideational dimension of how to organize such work, the analysis has shown that divergence has arisen as the idea has traveled. Overall two different types of evidence-producing organizations have been established. One type exemplified by the Cochrane Collaboration and the Campbell Collaboration is international aiming at producing globally valid knowledge. Another type is local—typically national—aiming at producing knowledge specifically relevant to national political as well as professional actors. In this connection different national agendas and different political and bureaucratic actors have shaped the organizational dimension of the idea.

Finally, also in relation to the methodological ideational dimension of how to carry through review practice, the analysis has shown that also in this respect divergence has arisen as the idea has traveled. A variety of approaches to review practice have been developed, some associated with a hierarchy of evidence thinking others with a typology of evidence thinking. Also a range of approaches to synthesizing has been developed. Methodological divergence has arisen as the idea has traveled out of the health field and into the fields of social welfare and education as well as it has traveled out of the United States and into the United Kingdom. In this connection different scientific paradigms and traditions have shaped the methodological dimension of the idea.

The pattern of divergence is illustrated in Figure 4 which combines the methodological dimension with the organizational dimension.

Combining these dimensions crystallizes four different approaches to review practice. One approach is to depart from the hierarchy of evidence having the ambition to produce universal evidence. This is the

Figure 4
A Typology of Evidence Thinking

Evidence thinking:	Hierarchy of evidence	Typology of evidence
Global: Evidence is universal	Global reviews including only RCTs (Cochrane, Campbell).	Global reviews including primary studies with a variety of designs.
Local: Evidence is contextual	a) National reviews including only RCTs (WWC). b) Results from contextual relevant RCT designed primary studies are more useful than global reviews.	National reviews including primary studies with a variety of designs (EPPI).

approach used by both the Cochrane and the Campbell Collaboration. A second approach is to depart from the hierarchy of evidence having the ambition to produce national evidence. This approach is used by WWC which in defiance of their contractual collaboration with Campbell only include results from primary studies conducted in the United States in their systematic reviews. At the Campbell Colloquium held in 2005 the above-mentioned Robert Boruch was asked why the approach of the WWC was national and not global. His answer was that this was due to considerations for educational practitioners. WWC did not expect U.S. teachers to accept knowledge from abroad as legitimate. WWC thus seems to compromise with their own methodological position in order to adapt to important stakeholders.

On the Nordic scene an interesting discussion among hierarchy of evidence disciples has been continuing as to the value of global versus local evidence in the field of social work. The discussion has been brought about by the publication of a systematic review on multi-systemic therapy. Outside the medical field RCT designed studies most often compare the effects of a new intervention with treatment as usual. When results from several primary studies conducted in countries with very different social welfare systems, and thus different types of treatment as typical are synthesized, intervention effects measured on different scales so to speak are weighted together. In continuation of the methodological problems raised by this it has been argued that it is more useful to synthesize results from contextual related primary studies—or even use the results from a single contextual relevant RCT study—than to have an ambition to synthesize universal evidence.

A third approach is to depart from the typology of evidence having the ambition to produce national evidence. Many of the EPPI reviews in UK are examples of this.

Finally, a fourth approach is to depart from the typology of evidence having the ambition to produce universal evidence. We have found no evidence-producing organizations subscribing to this approach. However, the Danish Clearinghouse in its policy for review practice opens up for this opportunity. The Danish Clearinghouse states that it will disseminate international reviews and assess these as to clear out whether they are relevant in a Danish and Scandinavian context. The strategy thus seems to be to produce and spread both global and local evidence according to assessments of different types of interventions.

Today, most evidence-producing organizations seem either to go for universal or national evidence. It can, however, be discussed whether these units of analysis are the only relevant ones. It is surprising that we only to a very limited extent find evidence-producing organizations with for example European or regional ambitions.

Reflecting on the more general mechanisms behind the upcoming of evidence-producing organizations during the last 15-20 years our assumption based on data from our study is that it has been a merry alliance between researchers, professionals within the specific policy areas and government with international organizations as facilitators. Researchers have seen a new product (systematic reviews) that can be sold as well as a means for providing academic prestige. Professionals have recognized a need for ways of coping with the increasing amount of knowledge supply. Governments have the similar need, but also a need for ways of managing knowledge for decision making that has a high potential of legitimacy. International organizations such as OECD, but also international professional organizations, have been facilitating the spread of the evidence idea, and providing platforms for sharing knowledge of and even imposing methods and dissemination of systematic reviews on national governments. The national governments have been using international organizations as intelligence and critical observers, and international organizations have functioned as brokers between government representatives, the research community and professional networks in relation to strategies and discourse on evidence.

Notes

1.   The research project has received funding from the Danish Social Science Research Council. We thank Yosef Bhatti and Leo Milgrom for an extensive effort in connection with the searching, processing and analysis of relevant documentary data.

References

Boaz, A., D. Shelby, D. Denyer, M. Egan, A. Harden, D.R. Jones, R. Pawson and D. Tranfield (2006). "A multitude of syntheses: a comparison of five approaches from diverse policy fields." *Evidence and Policy* 2,4: 479-502.

Cabinet Office (1999). *Modernising government [online]*. England: The Stationary Office.

http://www.archive.official-documents.co.uk/document/cm43/4310/4310.htm (accessed February 13, 2007).

Cabinet Office Performance and Innovation Unit (2001). *Better policy delivery and design: A discussion paper*. London: Cabinet Office.

Chalmers, I. (2003a). "Fisher and Bradford Hill: theory and pragmatism?" *International Journal of Epidemiology* 32: 922-924.

Chalmers, I. (2003b). "Trying to do more good than harm in policy and practice: The role of rigorous, transparent, up-to-date evaluations." *Annals of the American Academy of Political and Social Science* 589: 22-40.

Clarke, A. (2006). "Evidence-based evaluation in different professional domains: Similarities, differences and challenges." In I.F. Shaw, J.C. Greene and M.M. Mark, *The Sage Handbook of Evaluation*, London: Sage.

Czarniawska, B. and B. Joerges (1996). "Travel of ideas." In B. Czarniawska and G. Sevon (eds.), *Translating organizational change*, New York: Walter de Gruyter: 13-48.

DiMaggio, P.J. and W.W. Powell (1991). "The iron cage revisited: Institutional isomorphism and collective rationality in organizational fields." In W.W. Powell and P.J. DiMaggio (eds.), *The new institutionalism in organizational analysis*, Chicago: University of Chicago Press.

Dixon-Woods, M., R.L. Shaw, S. Agarwall and J.A. Smith (2004). "The problem of appraising qualitative research." *Quality and Safety in Health Care* 13: 223-225.

Farrington, D.P. (2003). "Methodological quality standards for evaluation research." *Annals of the American Academy of Political and Social Science* 587: 49-68.

Gray, J.A.M. (1997). *Evidence-based healthcare*. London: Churchill Livingstone.

Hansen, H.F. and O. Rieper (in print). *Metodedebatten om evidens*. Copenhagen: AKF Forlaget.

House, E.R. (2004). *Democracy and evaluation*. Keynote speech at European Evaluation Society Conference in Berlin.

Launsø, L. and D.E. Gannik (2000). "The need for revision of medical research designs." In D.E. Gannik and L. Launsø (eds.), *Disease, knowledge and society*, Frederiksberg: Samfundslitteratur: 243-261.

Macdonald, G. (2000). "Social care: rhetoric and reality." In H.T.O. Davies, S. Nutley and P.C. Smith (eds.), *What Works? Evidence-based policy and practice in public services*, Bristol: the Policy Press.

OECD (Organisation for Economic Co-operation and Development) (1995). *Educational research and development: Trends, issues and challenges*. OECD.

OECD (Organisation for Economic Co-operation and Development) and IES (International Evaluation Society, Coalition for Evidence-Based Policy) (2004). *OECD-U.S. Meeting on Evidence-Based Policy Research in Education*. Forum Proceedings, April 19-20.

Pawson, R. (2006). *Evidence-based policy. A realist perspective.* London: SAGE Publications, Ltd.

Petrosino, A., R.F. Boruch, H. Soydan, L. Duggan, and J. Sanches-Meca (2001). "Meeting the challenge of evidence-based policy: The Campbell Collaboration." *The Annals of the American Academy of Political and Social Sciences* 578: 14-34.

Petticrew, M. and H. Roberts (2003). "Evidence, hierarchies and typologies: Horses for courses." *Journal of Epidemiology and Community Health* 57: 527-9.

Petticrew, M. and H. Roberts (2006). *Systematic reviews in the Social Sciences.* Malden: Blackwell Publishing.

Radaelli, C.M. (2005). "Diffusion without convergence: how political context shapes the adoption of regulatory impact assessment." *Journal of Public Policy* 12,5: 924-943.

Røvik, K.A. (1998). *Moderne organisasjoner.* Bergen: Fagbokforlaget.

Sahlin-Anderson, K. (1996). "Imitating by editing success: The construction of organizational fields." In B. Czarniawska and G. Sevon (eds.), *Translating organizational change*, New York: Walter de Gruyter: 69-92.

Schneider, E. (2002). "The Campbell Collaboration: Preparing, maintaining, and promoting the accessibility of systematic reviews of the effects of social and educational policies and practices." *Hypothesis—the newsletter of research* 16,3.

Shadish, W.R., S. Chacón-Moscoco and J. Sánches-Meca (2005). "Evidence-based decision making: Enhancing systematic reviews of program evaluation results in Europe." *Evaluation: the international journal of theory, research and practice* 11,1: 95-109.

Solesbury, W. (2001). *Evidence Based Policy: Whence it Came and Where it's Going.* Working paper 1. ESRC UK Centre for Evidence Based Policy and Practice, Queen Mary, University of London.

Part II

Generating Evidence

4

What Counts as "Sufficient Evidence" and a "Compelling Argument" in Evaluation?

Tom Ling

> Out of the air a voice without a face
> Proved by statistics that some cause was just
> In tones as dry and level as the place:
> No one was cheered and nothing was discussed;
> Column by column in a cloud of dust
> They marched away enduring a belief
> Whose logic brought them, somewhere else, to grief.
>
> W.H. Auden
> "The Shield of Achilles"

Introduction

We live in an age awash with evidence; in public policy, management and professional practice there are ever more facts available to decision-makers. Yet neither do decision-makers appear more confident in their choices nor do the intended beneficiaries of these decisions appear better served.[1] It has long been recognized that facts need to be organized, weighed against each other for validity and reliability, and structured to create an evidence-base. During the modern age this has been accomplished through the sequestration and professionalization of knowledge creation and the expansion of organizational life. The former has created a cadre of "experts" (Stilgoe, Irwin and Jones, 2006), academies and research organizations who trade in the quality of the evidence they generate within rules that they claim responsibility for policing. The latter has created an increasing number of organizations whose reach now extends into areas of life which in the past were informal and structured through traditions and social mores rather than organizational forms.

There is already interest in knowledge as power, technocratic expertise as a basis for legitimizing action, and the role of new forms of evidence (such as statistics, profit and loss accounts and so forth) associated with key figures in the social sciences such as Weber, Foucault and Giddens. This chapter has a more specific aim of understanding and explaining how the evidence movement might reshape the way evidence is produced, assessed and used in many of the most important decisions in our private, social, and political lives.

Understanding the need that modern organizations have for evidence is an important part of this discussion. Most modern organizations are hungry for formal evidence; in the knowledge economy prospecting for facts has replaced prospecting for gold. Getting the right sort—and right balance—of information is seen to be crucial to supporting successful decision-making (see Kaplan and Norton, 2001, 2006). It may also be the case (but this would be the subject for a different paper) that more recent organizations are more networked, less prone to trust professionals and bureaucrats, and seek to be more responsive to their environment; for these reasons, on average, they may be more "evidence-hungry" than their predecessors. And whether or not organizations are more evidence-hungry, there are certainly more of them. Consider, to take a random example, the 3,500 NGOs registered in Uganda in 2000 alone or the five-fold increase in companies registered in California 1960-2001 (Lyman, Meyer, and Hwang, 2006). Of the 3,500 additional NGOs in Uganda most, if not all, would have monitoring and evaluation frameworks and would assess needs and impacts. The consequences of the growth both professional knowledge creators and of organizational life can be seen in the production of ever more evidence about ever more aspects of social and political life. According to one report, the amount of new information stored in different media doubled in the three years up to 2002 (Lyman and Varian, 2003). In the sense that our collective capacity to generate evidence was outstripping our capacity to use it, the Evidence Based Movement was a movement waiting to happen.

If there was an increase in the quantity of evidence being generated, there was also a change in the quality of evidence that was being used. Eraut (2000) distinguishes between propositional, or codified, knowledge and non-propositional, or personal, knowledge. The former draws upon formal knowledge generated from research while the latter draw upon the experience of the practitioner and their ability to perform in routine settings. Recent years have seen the privileging of the former over the latter. For example, healthcare systems across the world have invested heavily in systems to disseminate and promote adherence to research evidence but invested less in systems to improve the use of tacit and personal knowledge.

However, the evidence explosion has not led to a sense of greater accountability nor has it obviously led to more organizational learning. In part this apparent paradox might be because of the psychology of choice; faced with ever more, and more disparate, information, individuals may become less able to select and less able to aggregate evidence into usable knowledge. This is parallel to the "paradox of choice" identified by Schwartz (2004) where being offered choices between a growing number of incommensurate options becomes no choice at all. Ever more incommensurate pieces of evidence destabilize efforts to make sense of the world, reinforcing the sense that everything is very complex without providing the resources to manage this complexity.

However, it is not just a problem of individual psychology. It is a feature of communities of knowledge creation and communities of practice that they equip their members to handle this complexity by providing shorthand routes or guides to allow evidence to emerge from the myriad world of dull facts. This is at heart a social process. Even intangibles such as "leadership," "legitimacy," or "power" are constructed into a meaningful source of evidence that both explains and reveals the world. These communities, what will be called in this chapter "epistemic communities" (a term explained below), provide supports for establishing what counts as evidence, how to weigh one piece of evidence against another, how to aggregate data and when judgments may be justified—even if that is a provisional one. Discussing the diverse ways that evidence is created, ranked and used, Davies and colleagues comment: "These differences may arise for a number of reasons: the nature of the service; the costs of providing evidence; the capacity and culture of the research community; the attitudes and prejudices of practitioners; the attitudes and prejudices of citizens and policymakers; or historical accident" (Davies, Nutley and Smith, 2000: 351). The kinds of evidence contributing to an epistemic community have been well described by Rycroft-Malone and colleagues (although not using the term "epistemic communities") as including: formal research; professional knowledge and experience; local data and information; and user experiences and preferences (Rycroft-Malone et al., 2004). They argue that in practice these elements are melded in individual decisions made close to the individual service user or client.

Communities of practitioners provide decision-makers with the resources to avoid the unmanageability and disruption of extreme epistemological skepticism—where every claim is subject to equally extreme doubt—by establishing certain sources, certain methodologies, certain experts or certain practices as the arbiters of what we can accept and what requires further attention. These provide the paradigms within which practice and progress can be managed. However, it brings with it the risk of epistemological arrogance—where only knowledge conforming to the norms of the community can be accepted.

These general problems facing all of us in modernity have particular implications for evaluators. Evaluators sit at the intersection between the professionalization of knowledge creation and the expansion of organizational life. Because they seek to re-conceptualize and re-categorize organizations, programs or other interventions and to then redescribe them (often in terms of purpose, inputs, processes, outputs and outcomes, or similar frameworks), they do not always sit comfortably with the assumptions and claims of the epistemic communities they encounter in the course of their evaluations.

Evaluators have a view (indeed many views) on what counts as evidence, and how evidence is to be presented and weighed. Elsewhere, rules governing this use of evidence might be more or less formal but different social groups (such as scientists, lawyers, clinicians) exhibit there own definitions and practices. Often this is associated with the emergence of professions with, for example, lawyers claiming the right to monopolize what counts as legal "facts," and medical doctors identify medical evidence. The power to define what constitutes evidence can often provide a basis of wider social power. We will use the term "epistemic community" to characterize a group of practitioners, policymakers and clients who share an understanding the nature of evidence. This follows the observations of Kuhn (1962), but also the more recent discussions on the sociology and anthropology of science associated with, for example, Michel Callon (1987, 1991) and John Law (2008, 1999). Evaluators make an unusual community because they form a "trans-discipline" both sitting above and depending upon the evidence produced by other professions and disciplines. This also provides them with the opportunity to play a positive role in this debate.

Epistemic communities may be understood from at least three standpoints relevant to evaluation. First, there is a sociology of epistemic communities that explores their emergence in terms of social agency and process. Such an approach might consider, for example, the sequestration of medical knowledge by the medical profession during the nineteenth century to the detriment of healers, herbalists, and other traditional providers of health care. From this standpoint, evidence-based decision-making should be seen as a social process involving the production and grading of evidence, on the one hand, and the application of this evidence to particular decisions in policy or practice. Key questions include who and what is included, what rules are applied, how is dissonance managed, and how is consensus built. Organizationally, financially and socially, evaluators often work at the very heart of this process. Second, there is an epistemological approach that would interrogate the philosophical bases for a community's epistemological claims. Epistemic communities make epistemological claims (implicitly or explicitly) and evaluators—a trans-discipline making no unique epistemological claims of its own— must inevitably engage with these claims. Considerable philosophical effort has been put into interrogating the epistemological claims of the evidence-based movement and there exists a rich diversity of opinion. Third there is a practitioner dimension where members of the community seek to become expert in using evidence appropriately (according to the rules of their own community). Practitioners not only provide much of the evidence used in an evaluation but they are also often the target group for evaluators wishing support improvements in services by changing professional practice and consequently evaluators may wish to equip themselves with a sound understanding of that community.

Epistemic communities, it should be stressed, are often uneven, over-lapping and contradictory. The sociological definition of the medical profession might fit uncomfortably with some practitioners' views—for example, some practitioners are more questioning of the power of the medical profession and resist the epistemological claims of the so-called "medical model." Even so, the evidence movement is more challenging for some epistemic communities than others and, within epistemic communities, for some practitioners more than others. Therefore evaluators do not enter epistemic communities that are fully formed, internally cohesive, and entirely solidaristic. But they do enter worlds where there already exists a set of pre-existing assumptions about what counts as evidence, what hierarchies there are, and what a compelling argument looks like.

I wish to argue that in all three dimensions—sociological, episte-mological and practitioner—that elements of the so-called "evidence movement" have encouraged an epistemic arrogance, an inability to weigh evidence, and an exaggerated confidence in global experts over local practitioners. The argument to be developed here is parallel to (but at this stage less well documented) Elster's conception of "local justice" (Elster, 1992). Elster argues that, in examining how particular institutions allocate scarce resources (access to transplant organs, for example) or necessary burdens (drafting to the armed forces, say) they make decisions involving efficiency and equity in ways that reflect the particular history and setting of that organization. In this way they seek to arrive at just and fair decisions but these notions of justice and fairness are far removed, yet more pragmatic, than the "global" theories of justice beloved by philosophers and theorists. In weighing up how to allocate scarce transplant organs, for example, questions of tissue type, how long patients have been on the waiting list and so on are invoked in order to arrive at an acceptable judgment. In the parallel argument suggested here, these "local" settings also provide the context for selecting evidence, collecting new data, and weighing this evidence. Because this process is often closely connected to a sense of what is fair and just (in the local pragmatic sense) the "global" claims of the evidence movement may have little traction or even provoke hostility where this does not fit comfortably with evidence from practitioner experience, user views, or local capacities.

As evaluators, we have no particular interest in privileging the local over the global. Indeed, given local power asymmetries there are good reasons to anticipate that aggregated decisions of "local justice" can lead to "global injustice." However, neither should we necessarily privilege the global claims over the local. Indeed, the global claims may lack the granularity needed to support, or to understand, how evidence is used in practice, how this is embedded in assumptions of fairness and efficiency, and how the smooth running of local practices depend upon shared, path-dependent knowledge. "Local evidence" might be local in the sense that it is based upon evidence only known locally but it might also be local in the sense that it reflects some historic compromise about whose voices should be heard and what was considered relevant information in the debate. These pragmatic compromises might bring their own benefits and these should not be dismissed lightly.

Therefore, evaluators might be well advised to adopt a more humble view of their work and a more cautious interpretation of what constitutes "sufficient evidence" and a "compelling argument." This is emphatically not an argument against evaluation or using evidence, but it is an argument for creating an environment in which evaluation can add value and satisfy expectations at the local as well global level. It argues for a richer view of what counts as evidence and a more restrained account of what constitutes a compelling argument.

Epistemic Communities

The practice of producing and using evidence intersects with decision-making in distinct ways according to their social settings. In criminal investigation, for example, the development of forensic science introduced a whole set of practices designed to connect suspects to the time and place of the crime. When evidence is produced through, for example, forensic pathology, it cannot be used "raw" in the judicial process. Rather it is presented along with the a description of the processes used to produce it, the qualifications of those doing it, the likelihood of the evidence being misleading and so forth. Alongside the technology linked to forensic science there developed a whole routine of how crime scenes should be managed. In law what can be counted as admissible evidence is a matter of law itself (as is who or what can bring evidence, sue or be sued, and what constitutes a legal subject). The use of expert witnesses is an interesting (and increasingly controversial) aspect of the legal process such that expert judgment is not treated as legal judgment but does count as a privileged part of the evidence base. In national security and intelligence work we are (for obvious reasons) often unable to judge how evidence is used but anecdotally it appears that weight is attached to the source of the story and these sources are somehow "ranked" through an informal process which involves arbiters in somehow valuing some sources above others. Meanwhile in medicine, as described in Chapter 7 of this volume, there is a formal hierarchy of a different sort which ranks the methodology used to assess the comparative value of the evidence produced. Across all of these areas there has been substantial investment of effort and money to create infrastructures which are intended to ensure that decisions are based upon the best available evidence (indeed, the term "quality" has come to be associated with the idea that the service so organized as to ensure that the gap between the best evidence available and the actual decisions taken is minimized).

The experience of this consequential explosion of evidence is amplified by electronic sources available at the stroke of a mouse. Yet evaluations do not appear to have become any easier and, indeed, the pages of international evaluation journals suggest a community of practitioners in energetic debate rather than one which is quietly confident that the accumulation of evidence will render up the truth. There are two important reasons for this—the first to do with the impartiality of the evidence collected and the second to do with how to connect evidence to evaluation findings. The first is that most routine evidence is produced and collected by social groups with their own interests and the evidence they collect reflects such interests. For example, far more data in health and medical science is concerned with identifying potential new treatments rather than explaining why physicians so often fail to use the evidence that already exists. Evaluators must then either mine data collected for one purpose with a different purpose in mind or collect new data. The "pre-formed" nature of available evidence is therefore a potential barrier to independent analysis. The second barrier is that evaluation asks questions which in most non-trivial cases can only be answered with qualifications and caveats.

Finally, epistemic communities provide the basis for thinking about more that just evidence. They also provide the framework for how specific communities in their local setting think about justice. What is regarded as fair in the treatment of clients, colleagues, managers, and public reflect not only more "global" concepts such as "equality" but also mediate these universal terms through the prism of local knowledge. Of necessity, when we apply global principles we do so through the prism of the local circumstances in which we live and perform. Thus Barker argues that:

> The notion that we should—or perhaps even could—base our practice on "generalizable evidence" demolishes our traditional practice. Such worldviews urge us to swap our ideas of crafting care around the unique complexity of the individual, for a generalization about what worked for most people in a study (Barker, 2000: 332).

However, although this view explains some of the resistance to "generalized" evidence, and although evaluators should take account of the local contexts in which evidence is used, they are also required to draw upon a balanced set of formal and tacit evidence about organizational performance. There are some ideas about how to achieve this found in organizational tools such as balanced scorecards and dashboards.

Weaving Sense of Complex Facts: the Examples of the Scorecard and the Dashboard

Evaluators face the task of drawing upon disparate sources of evidence within an organization. To this they will add the data produced by their own research. They must then seek to come to some overall judgment but this is constrained by the incommensurate nature of much of the evidence; in some dimensions the program might be performing less well than in others. These evidence bases cannot always be aggregated and neither can they easily be placed in a hierarchy. One solution to this is to draw upon the ideas around multi-criteria decision-making by making apparent the different criteria that matter to an organization or program and its stakeholders and to collect evidence that casts light on performance within each of these criteria. Two important examples of such devices used in the United Kingdom are scorecards and dashboards.

Scorecards and dashboards represent an attempt by organizations, faced with the profusion of evidence, to identify the "minimum data set" that might allow them to remain focused on the strategic goals and to monitor performance against the things that really matter. In the United Kingdom, the sense that simply collecting more and more data is eventually unhelpful has created a keen interest in how to aggregate and balance evidence. These benefits are articulated clearly in *Choosing the right FABRIC*, published jointly by the Treasury, Cabinet Office, National Audit Office, Audit Commission and Office for National Statistics in 2001. (HM Treasury, 2001). The intended benefits may be summarized as: "Performance information is key to effective management" and "Externally, performance information allows effective accountability" (HM Treasury, 2001: 4). In common with most approaches to performance evaluation, the report recommends collecting data around the logical flow of delivery: resources, inputs, outputs and outcomes. They are a pragmatic way of presenting evidence to support judgments in a timely and focused way. They are not, therefore, innocent tools.

The years following the publication of FABRIC in 2001 saw not only a growth in the quantity of performance information but also a growing recognition of the possible perverse consequences of such information (and indeed these are flagged up in the report) (HM Treasury, 2001:18). The Public Administration's Select Committee 2003 report *On target? Government by measurement* (Select Committee, 2003) the Royal Statistical Society (Bird et al., 2003) and others highlighted both the managerial and technical limitations and the challenges of performance data. These anxieties reiterate concerns expressed since the 1950s (if not earlier) and identified again in the discussion by Smith (Smith, 1995) of the dysfunctional effects of the publication of performance data.[2] This has encouraged a view that, for Key Performance Indicators (KPIs), more is not always better and that, like the organizations they quantify, KPIs should be lean. It also encouraged a view that micro-managing through performance indicators would, in many contexts, only encourage perverse outcomes. There is also a sense that performance indicators may not always be effective instruments of control and as Goodhart's Law famously states: "any observed statistical regularity will tend to collapse once pressure is placed upon it for control purposes" (Goodhart, 1975).

What we are witnessing in this debate is that the identification of good performance data and their aggregation as KPIs is faces challenging demands. One demand (articulated by the Royal Statistical Society in the United Kingdom) is that it should be technically competent. Another is that KPIs should tell elected politicians what they need to know to entrench accountability (represented by the Select Committee). A third is that we should stand back and look at the consequences of KPIs on society more widely (represented by Smith and like-minded colleagues).

In common with many other organizations seeking to develop key performance indicators, Kaplan and Norton's "balanced scorecard" (Kaplan and Norton, 1992) was seen to be a promising place to start. The Balanced Scorecard was originally developed in a business context, in response to performance measurements that were solely focused on financial or operational data. It was intended to provide a more sophisticated performance management system that encouraged managers to consider their business from four different perspectives: financial, internal process, innovation and learning, and customer (Kaplan and Norton, 1992: 72) Therefore, the Balanced Scorecard brought "balance" by encouraging companies to incorporate into their performance management intangibles such as customer satisfaction, employee goodwill, staff training, brand recognition, as well as established financial metrics. For example, recognizing the need for a "multi-attribute" approach, there has been "a gradual increase in awareness of the Balanced Scorecard among the R&D management community at both theoretical and practitioner levels" (Osama, 2005: 3).

However, constructing KPIs, even in the context of the Balanced Scorecard, runs the constant risk of alienating local practitioners, for the reasons suggested above. Similarly guidelines, targets, standards and evidence-based protocols can be equally alienating, often provoking local anger and frustration. The role of evaluation in this context could be crucial. Rather than evaluating against the global evidence, on the one hand, or simply accepting and evaluating against the framework of local evidence, on the other hand, evaluations can function to "translate" between these levels and to challenge each against the evidence produced by the other.

This suggests a significantly different role for the evaluator; a role requiring additional skills and practices. Evaluators must not only be adept at collecting and using evidence to evaluate performance against a particular criterion, but they must also understand the criteria that matter to the different stakeholders. They must recognize how different "epistemic communities" not only generate evidence differently but also how their distinct values shape this and how they have arrived at local compromises with others in the process of maintaining the program or organization being evaluated. The balanced scorecard or dashboard is less an objective measurement and more the result of complex social processes. Some of the qualities an evaluator might hope to bring to this process include: transparency, inclusiveness, reliability and explicitness. But in addition, there are the social skills necessary for facilitating the inevitable dialogues that the evaluation will generate. It may not be sufficient to disrupt complex social processes without taking some responsibility for minimizing the harm that might flow from this and maximize opportunities for social benefit. This might be seen as the diplomacy of speaking truth to power.

Concluding Remarks

Auden was quoted at the start of this chapter writing:

> Out of the air a voice without a face
> Proved by statistics that some cause was just
> In tones as dry and level as the place:
> No one was cheered and nothing was discussed

This chapter has argued that the EBM may encourage a false confidence in easy proofs that "some cause is just." There are neither good philosophical nor practical reasons for such confidence. Rather, the EBM offers a partial view of how the relationship between evidence and decision-making should be socially managed. This is an unnecessarily constricted view which misses the inevitable social dimensions of evidence-based decision making involving multiple epistemic communities. Without simply privileging the local over the particular, or the subjective over the objective, the evaluator might hope to clarify and communicate these multiple criteria and support evidence-informed evaluations of each of them and support a deliberation about performance across multiple dimensions. The chapter argues against an outcome where "no one was cheered and nothing was discussed."

The argument draws upon wider literature linking the sociology of knowledge and the study of modern organizations and uses these to inform a discussion about the intersection of evaluation and the Evidence Based Movement. At heart, it is a plea for evaluators to regard the evidence they use with suspicion and care rather than to trust everything near the top of the hierarchy of knowledge and distrust everything near the bottom. It is firmly not, however, a plea to invert this hierarchy and nor is it a postmodern claim that all truth is discursively constituted and that all discourses are equal.

Organizational life and knowledge creation have come together in late modernity to create both a demand and a supply for evidence. The Evidence Based Movement is both an expression of this and a driver. However, evidence is collected and weighed in different ways according to the traditions and requirements of different epistemic communities. Furthermore, organizations also actively sift, balance, and manage this information to produce balanced scorecards, metrics, targets, and dashboards (among other things). Consequently, information within organizations and policy areas may have been substantially sifted before it is taken up and used by evaluators.

Faced with this difficulty, evaluators may seek to assert the "superiority" of their own analysis. This carries the risk that those being evaluated fail to recognize the description offered by the evaluators. The dissonance between these may partly explain why not all evaluations produce changes in policy and practice. And this dissonance is not simply a matter of ignoring evidence. It can also be a dissonance about values and matters of justice, purpose, and fairness.

Alternatively, evaluators may adopt a different approach. This would involve a process of deliberative evaluation in which the evaluator systematically exposes the multiple criteria used by stakeholders and establishes the evidence appropriate to assessing the achievement of objectives based on these criteria. This may avoid simplistic evaluations and erroneous conclusions "Whose logic brought them, somewhere else, to grief."

Notes

1.   "Evidence" is taken to mean systematically collected and analyzed data used to support choices; "facts" claim to reveal discrete and particular aspects of the real world.
2.   These might be summarized as: tunnel vision; sub-optimization for the whole system; myopia; measure fixation; misrepresentation; misinterpretation; gaming; ossification.

References

Barker, P. (2000). "Reflections on caring as a virtue ethic within an evidence-based culture." *International Journal of Nursing Studies* 37: 329-336.
Bird, S.M., D. Cox, V.T. Farwell, K. Goldstein, T. Holt and P. Smith, P. (2003). *RSS Working Party on performance monitoring in the public services*. London: Royal Statistical Society.

Callon, M. (1987). "Society in the making: The study of technology as a tool for sociological analysis." Pp. 83-103 in W. Bijker et al. (eds.), *The social construction of technical systems: New directions in the sociology and history of technology*, London: MIT Press.

Callon, M. (1991). "Techno-economic networks and irreversibility." In J. Law (ed.), *A sociology of monsters: Essays on power, technology and domination*, London: Routledge.

Davies, H., S. Nutley and P. Smith (2000) "Learning from the past, prospects for the future." In H. Davies, H., S. Nutley and P. Smith, *What Works? Evidence-based policy and practice in public services,* Bristol: The Policy Press: 351.

Drori, G.S., W. Meyer and H. Hwang (2006) "Introduction." In G.S. Drori, W. Meyer and W. Hwang (eds.) *Globalization and organization.* OUP.

Elster, J. (1992) *Local justice: How institutions allocate scarce goods and necessary burdens.* New York: Russell Sage Foundation.

Eraut, M. (2000). "Non-formal learning and tacit knowledge in professional work." *British Journal of Educational Psychology* 70: 113-136.

Goodhart, C.A.E. (1975). *Monetary relationships: A view from Threadneedle Street.* Papers on Monetary Economics, Vol. 1, Reserve Bank of Australia.

HM Treasury, Cabinet Office, National Audit Office, Audit Commission and Office for National Statistics (2001). *Choosing the right FABRICA framework for performance information.* London: TSO

Kaplan, R. and D. Norton (1992). "The balanced scorecard: Measures that drive performance." *Harvard Business Review*, January-February: 71-19.

Kaplan, R. and D. Norton (1996). *The balanced scorecard.* Boston, MA: Harvard Business School Press.

Kaplan, S.R. & Norton P.D. (2001). *The strategy focused organization— how balanced scorecard companies thrive in new business environment.* Boston: Harvard Business School Press.

Kaplan, S.R. & Norton P.D. (2006). *Alignment —Using the balanced scorecard to create corporate synergies.* Boston: Harvard Business School Press.

Kuhn, T. (1962). *The structure of scientific revolutions.* Chicago: University of Chicago Press.

Law, J. (2008). "Actor-network theory and after." In B.S. Turner (ed.), *The new Blackwell companion to social theory* 3rd Edition, Oxford: Blackwell.

Law, J. and Hassard, J. (eds.) (1999). *Actor network theory and after.* Oxford: Blackwell and Sociological Review.

Lyman, P. and H. Varian (2003). "How much information." As of August 2007: http://www2.sims.berkeley.edu/research/projects/how-much-info-2003/

Osama, A. (2005). "Using multi-attribute strategy & performance architectures." In *R&D organizations: The case of balanced scorecards*, Santa Monica CA: Frederick S. Pardee RAND Graduate School: 3.

Rycroft-Malone, J., K. Seers, A. Titchen, G. Harvey, A. Kitson, B. McCormack (2004). "What counts as evidence-based practice?" *Journal of Advanced Nursing* 47,1: 81-90.

Schwartz, B. (2004). *The paradox of choice.* London: HarperCollins.

Select Committee 2003 report *On target? Government by measurement.*

Smith, P. (1995). "On the unintended consequences of publishing performance data in the public sector." *International Journal of Public Administration*, 18,2&3: 277-310.

Stilgoe, J., A. Irwin and K. Jones (2006). *The received wisdom opening up expert advice.* London: DEMOS.

5

Evaluative Evidence in Multi-Level Interventions: the Case of the Global Environment Facility

Jos Vaessen and Rob van den Berg[1]

Introduction

In the current era of evidence-based policy and results-based management, politicians, decision makers and managers (as well as other stakeholders) demand reliable evidence on the effectiveness of policy interventions. As policy interventions are often multi-level (i.e., policies, programs, strategies comprising multiple intervention activities) the question of how to aggregate and synthesize evidence across intervention activities arises. Multi-level evaluative evidence refers to the evaluative output (data, analyses, reports) covering a particular group of interventions, spanning two or more administrative levels of interventions. This might include, for example, a group of projects constituting a program, or a group of programs (each consisting of individual projects) making up an agency's overall intervention strategy, or a group of projects executed by different agencies, making up a portfolio or program of a funding agency. A possible complication in the evaluation of a strategy or a program is that it typically involves more variability in terms of intervention design, context and outcome than a single intervention such as an individual project. Conversely, program evaluation could be facilitated by the availability of good evaluative evidence at the level of single intervention activities. These two aspects among other things distinguish multi-level evaluation from the evaluation of single intervention activities. Accordingly, specific methodologies for generating multi-level evaluative evidence have been developed over time. This chapter addresses some of the challenges these methodologies are facing.

The case of this chapter is the Global Environment Facility (GEF) or more precisely the GEF evaluation office. The GEF is an international funding mechanism aimed at protecting the global environment. The GEF is an example of a so-called Global and Regional Partnership Program (GRPP). These programs exhibit the following characteristics (IEG, 2007: xvi):

- different partners contribute and pool resources toward achieving agreed objectives;
- activities are often global, regional, or multi-country in scope;
- a new organization with a governance structure and management unit is established to fund these activities.

Multi-level interventions require evaluative evidence for the same reasons as single intervention evaluative evidence. First of all, organizations and especially GRPPs are accountable for their actions not only at the level of discrete interventions but also, and especially so giving the particular institutional and management characteristics of GRPPs, at portfolio and program level. The questions "are we doing things right" and even more so "are we doing the right things" require systematic evidence at program level supported by evidence at the level of discrete intervention activities in order to be answered. Second, learning about effectiveness across interventions is of crucial interest to GRPP stakeholders. In general, single intervention evaluations are limited in terms of their knowledge generation potential (Cook and Campbell, 1979). More specifically, there are limitations in terms of the internal validity (i.e., is there a causal relationship between changes in goal variables and intervention activities) as well as the external validity (i.e., to what extent is this relationship valid for a larger group of interventions or contexts). Consequently, there is a need to clearly qualify the findings generated by these evaluations, and correspondingly, to be cautious about the potential to generate knowledge about the performance and effectiveness of particular types of interventions. To different degrees the available multi-level evaluation approaches are helpful in reducing these threats to validity in order to generate robust evidence on what works and why across interventions.

The fact that multi-level evaluation approaches are well-positioned to contribute to knowledge repositories on interventions does not make the practical implementation of multi-level evaluations, and more particularly the realization of their evidence generation potential, any easier. Given the diversity in intervention contexts the challenge is to extract evaluative knowledge with some degree of generalizability while at the same time taking account of the specificities of local-specific contexts and implementation.

There are a number of key methodological challenges in multi-level evaluation contexts. A first challenge concerns the issue of attribution. This refers to the problem of establishing a causal link between intervention results and changes in target variables. In multi-level evaluation the attribution challenge at portfolio or program level is more complex than at the level of a single project. One of the main reasons is the variability in types of interventions that a program encompasses. How should evaluators deal with divergent types of interventions comprising a portfolio? How can they draw useful lessons from diverse experiences? The second challenge is the issue of alignment. Regardless of the types of intervention, the question arises whether the data produced at the level of individual interventions (projects) are adequate (i.e., in the sense of the higher-level portfolio's objectives) and sufficient for the evaluative analysis at portfolio level. If not, what is the scope for expanding the information base during the multi-level portfolio evaluation? And third, the issue of aggregation itself. Are there any structures (e.g., information systems) or procedures in place that systematize data from single interventions to the portfolio level? In what ways can evaluators meaningfully combine data from individual interventions into higher-level analyses and consequently produce some type of statement on merit and worth and/or draw meaningful lessons across interventions (see also White, 2003; Yang et al., 2004)?

The purpose of this chapter is to assess the methodological basis underlying the evaluative evidence that is currently generated by the GEF Evaluation Office. We ask the following questions. First of all: How does current evaluation practice in the GEF Evaluation Office compare to the range of available multi-level evaluation methodologies? Second, given the methodologies that are available, what would be worthwhile and interesting ways to further improve the quality of evaluative evidence produced by the GEF Evaluation Office? And, more particularly, what can be learned from new developments in the field of evaluative synthesis approaches? Answering these questions is not merely an exercise of identifying "gaps" which accordingly could be "closed" by merely adding new methodologies to the repertoire of the office. In addition, while some methodologies are considered to be more "rigorous" than others, with built-in validity checks, the choice for applying more of these types of methodologies is not a straightforward one. Evaluation within an international and complex network organization such as the GEF is more difficult to organize and streamline than in most organizations. Moreover, the topic of environmental benefits is often difficult to understand and assess from a policy perspective. These issues pose restrictions as to what types of methodologies are to be preferred.

The chapter is organized as follows. First, we will quick-step through the landscape of different multi-level evaluation methodologies used to aggregate lessons across levels to allow evaluative judgments of merit and worth as well as learning across interventions within a program or portfolio of interventions. Broadly, one can differentiate between two groups of methodologies: methodologies that involve primary data collection at the level of discrete evaluations (e.g., projects) and methodologies that rely on secondary data only.[2] The latter group of methodologies refers to a diversity of synthesis approaches. Recent developments regarding these methodologies have led to promising new avenues to generate evidence on what works and why across interventions (see for example Pawson, 2002a; see also Hansen and Rieper, this volume). Subsequently, we discuss the current state of application of multi-level evaluation approaches applied by the GEF Evaluation Office. Comparing the methodological literature with evaluative practice enables us to explore avenues for further improving the methodologies underlying the multi-level evaluations managed by the GEF Evaluation Office. Finally, we present some conclusions.

The Global Environment Facility (GEF): A Brief Background

The UN conference on environment and development (UNCED) in 1991 in Rio de Janeiro led to several multilateral environmental agreements, such as the Convention for Biodiversity and the UN Framework Convention for Climate Change. At the same time, a funding mechanism for the actions coming out of UNCED and the conventions was created: the Global Environment Facility (GEF). This mechanism is funded through replenishments on a four year basis. The latest and fourth replenishment amounted to US$3.15 billion. Since 1991, the GEF has provided grants for more than 1,300 projects in 140 countries. It aims to support both developing countries and countries with economies in transition to implement projects and programs that protect the global environment. GEF grants support projects in six focal areas: biodiversity, climate change, international waters, land degradation, the ozone layer, and persistent organic pollutants.

The GEF itself does not execute projects. The three founding members of the GEF: the United Nations Development Program, the United Nations Environmental Program and the World Bank act as Implementing Agencies of the GEF. In recent years, seven other agencies have become Executing Agencies of the GEF: the four regional multilateral Banks (Asian Development Bank, African Development Bank, European Bank for Reconstruction and Development, Inter-American Development Bank), the International Fund for Agricultural Development, the Food and Agricultural Organization of the UN and the UN Industrial Development Organization. These agencies form a network through which grants are distributed to recipient countries. At the core of the network, a relatively small Secretariat supports the GEF Council, which governs the GEF on behalf of the Assembly of GEF members, which only meets at the occasion of replenishments, every three or four years.

In the first years of the GEF, monitoring and evaluation was expected to be taken care of by the Implementing Agencies. In 1994, the GEF Council decided to create a Monitoring and Evaluation Unit in the GEF Secretariat, to enable cross-cutting monitoring and evaluation through the network. In 2003 the Council decided that this Monitoring and Evaluation Unit should become independent and report directly to the Council. In a few years the Unit transformed itself into the Evaluation Office of the GEF. The responsibility for monitoring was transferred to the GEF Secretariat.

The GEF Monitoring and Evaluation (M&E) policy describes the roles and tasks of the various partners in the GEF network. Monitoring is seen as a responsibility of management: it is first and foremost seen as a tool to keep activities on track. Evaluation is seen as an independent and impartial assessment of projects, programs and strategies. Evaluations should "provide evidence-based information that is credible, reliable, and useful, enabling the timely incorporation of findings, recommendations, and lessons into the decision-making process" (GEF, 2006a: 3).

Figure 1 provides an overview of the different levels of monitoring and evaluation in the GEF, as well as the different partners involved. The GEF Evaluation Office is responsible for corporate level evaluations. It currently has an annual budget of approximately $4 million and has a staff of eight evaluation professionals, as well as support staff including young evaluation professionals. Recent evaluations have been, amongst others:

- Local Benefits in Global Environmental Programs (2005)
- Evaluation of GEF support for biosafety (2006)
- Joint Evaluation of the GEF activity cycle and modalities (2006)
- Evaluation of Incremental Cost Analysis (2006)
- Country Portfolio Evaluations of Costa Rica (2006), The Philippines (2007), and Samoa (2007)
- Joint Evaluation of the Small Grants Program (2007)

Furthermore, an Annual Performance Report, of which the third was published in June 2007, combines a meta-evaluation of ex post project evaluations with assessments of factors impacting on the results of the GEF. In 2007, the first of a new series of annual reports was published on the impacts of the GEF, based on on-going impact evaluations. Other on-going evaluations are on the catalytic role of the GEF and on capacity development, as well as a new series of Country Portfolio Evaluations in Africa (Benin, Cameroon, Madagascar and South Africa).

Figure 1
Monitoring and Evaluation in the GEF

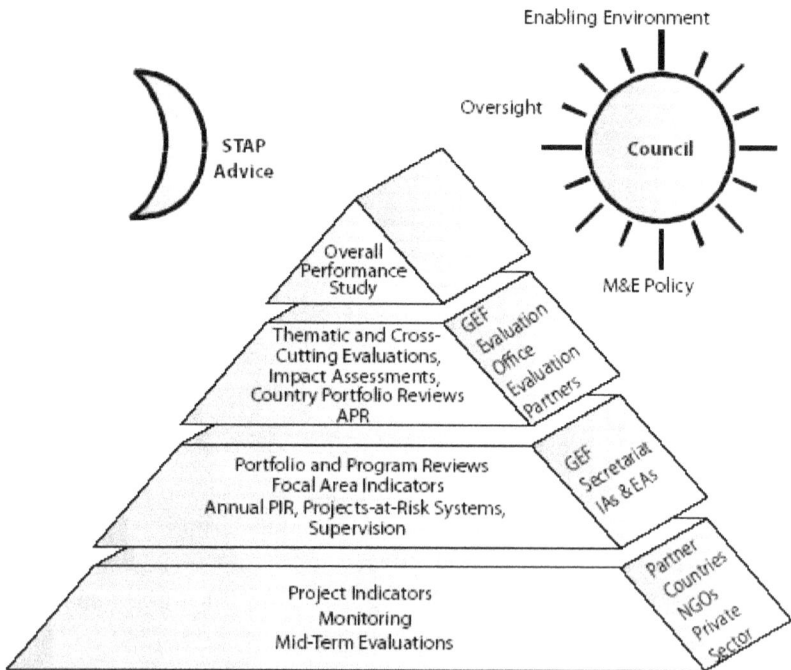

Source: GEF (2006a).

The evaluations of the GEF Evaluation Office are primarily directed to the GEF Council and are intended to inform policy decisions of the Council. However, they are also informing actions of the GEF Secretariat and the GEF Agencies. Furthermore, the recent development of Country Portfolio Evaluations has added recipient country governments as a new audience for these particular evaluations. All evaluations are published and available for all GEF partners and the Evaluation Office has a policy of actively promoting knowledge sharing.

In its evaluations, the Evaluation Office is tasked to include synthesis of evaluation results from "lower" level evaluations, undertaken at agency or project level. As shown in Figure 1, the evaluations of the Evaluation Office are all multi-level. They combine information generated at project level with data and analysis at the portfolio level including some type of synthesis approach as one of the methodological components. Furthermore, evaluations usually include field work at different levels as well: project visits and data gathering are often combined with surveys and stakeholder assessments at higher levels. With the exception of the Country Portfolio Evaluations, all evaluations of the Office take place in more than one country and often in more than one region, and typically involve more than one GEF agency. The Country Portfolio Evaluations take place in one country, but include projects from different focal areas which often are implemented in different geographic regions in the country, and implemented by different GEF agencies.[3]

The Landscape of Multi-Level Evaluative Evidence in a Nutshell

There is no existing generally accepted classification of multi-level evaluation methodologies. Broadly, multi-level evaluative evidence is generated through two groups of methodological approaches. A first group has been described by various authors under the banner of multi-site evaluations (Turpin and Sinacore, 1991; Straw and Herrell, 2002). Multi-site evaluations are by definition multi-level evaluations as they focus on programs encompassing multiple interventions implemented in different sites. Although often referred to as a family of methodologies, in what follows, and in line with the literature, we will use a somewhat more narrow definition of multi-site evaluations alongside several specific methodologies to address the issue of aggregation and cross-site evaluation of multiple interventions. A second group of techniques is associated with the term synthesis approaches and comprises a set of different methodologies that are used to assemble existing evaluative evidence at the level of single (site-specific) interventions as well as other sources of evidence in order to extract valid lessons on intervention effectiveness at higher levels of aggregation (e.g., Lipsey, 2000; Pawson, 2002a). The core of this group of techniques is the so-called systematic review, which refers to a set of procedural rules in order to collect and assess evaluative evidence towards an aggregate judgment. In addition, a number of more "soft" synthesis approaches can be listed under this banner (e.g., narrative reviews (see Pawson, 2002a)).

The following table developed originally by Worthen and Schmitz (1997) provides a useful overview of the two groups of approaches (see Table 1).[4]

The two groups of techniques mentioned earlier essentially differ in terms of when the decision is made to explicitly include multiple interventions (or intervention sites) in the empirical data collection phase of the intervention. Cells 1 and 3 in Table 1 cover evaluation approaches which primarily rely on primary data collection (during the evaluation process) at the level of single interventions. In contrast, the methodologies in cells 2 and 4 rely (almost) exclusively on existing data generated at single intervention level outside the direct framework of the multi-

Table 1
Basic Framework for Multi-level Evaluation

		Decision to use multiple sites is made:	
		Before evaluation begins	After evaluation data are collected
Program implementation in multiple sites is:	Controlled (the same in all sites)	1. Early decision to collect data from multiple sites; uniform (similar) implementation e.g., multi-center clinical trials	2. Analysis of existing site-level data; uniform (similar) implementation e.g., meta-analysis
	Uncontrolled (different across sites)	3. Early decision to collect data from multiple sites; variations in implementation, e.g., cluster evaluation	4. Analysis of existing site-level data; variations in implementation e.g., integrative review

Source: Adapted from Worthen and Schmitz (1997) and Potter (2005).

level evaluation. A second criterion differentiates between multi-level evaluations covering sets of (relatively) homogeneous interventions implemented in similar ways from programs or portfolios with variations in single intervention implementation or even different types of interventions altogether.

Cell 1

The multi-center clinical trial is a typical methodology of multi-level evaluation in which empirical data collection in a selection of homogenous intervention sites is systematically organized and coordinated. Basically it consists of a series of randomized controlled trials. The latter are experimental evaluations in which treatment is randomly assigned to a target group while a similar group not receiving the treatment is used as a control group. Consequently, changes in impact variables between the two groups can be traced back to the treatment as all other variables are assumed to be similar at group level. In the multi-center clinical trial sample size is increased and multiple sites are included in the experiment in order to strengthen the external validity of the findings. Control over all aspects of the evaluation is very tight in order to keep as many variables constant over the different sites. Applications are mostly found in the health sector (see Kraemer, 2000).

Cell 3

Cluster evaluation is a methodology that is especially useful for evaluating large-scale interventions addressing complex societal themes such as education, social service delivery and health promotion. Within a cluster of projects under evaluation, implementation among interventions may vary widely but single interventions are still linked in terms of common strategies, target populations or problems that are addressed (Worthen and Schmitz, 1997).

The approach was developed by the Kellogg Foundation in the nineties and since then has been taken up by other institutions. Four elements characterize cluster evaluation (Kellogg Foundation, 1991):

- it focuses on a group of projects in order to identify common issues and patterns;
- it focuses on "what happened" as well as "why did it happen";
- it is based on a collaborative process involving all relevant actors, including evaluators and individual project staff;
- project-specific information is confidential and not reported to the higher level; evaluators only report aggregate findings; this type of

confidentiality between evaluators and project staff induces a more open and collaborative environment.

Cluster evaluation is typically applied during program implementation (or during the planning stage) in close collaboration with stakeholders from all levels. Its purpose is, on the one hand, formative as evaluators in close collaboration with stakeholders at project level try to explore common issues as well as variations between sites. At the program level the evaluation's purpose can both be formative in terms of supporting planning processes as well as summative, i.e., judging what went wrong and why. A common question at the program level would be for example to explore the factors that in the different sites are associated with positive impacts. In general, the objective of cluster evaluation is not so much to prove as to improve, based on a shared understanding of why things are happening the way they do (Worthen and Schmitz, 1997).

Multi-site evaluation is another prominent multi-level evaluation methodology which also belongs in cell 3. In a way, multi-site evaluation constitutes the least well-defined methodology of the three we have introduced so far. The literature does not provide clear guidelines on what it exactly entails. Straw and Herrell (2002) use the term both as an overarching concept, i.e., including cluster evaluation and multi-center clinical trials, as well as a particular type of multi-level evaluation distinguishable from cluster evaluation and multi-center clinical trials. In this chapter we use the latter definition, the term multi-site evaluation referring to a particular (though rather flexible) methodological framework applicable to the evaluation of comprehensive multilevel programs addressing health, economic, environmental, or social issues.

Multi-site evaluation distinguishes itself from cluster evaluation in the sense that its primary purpose is summative, the assessment of merit and worth after the completion of the intervention. In addition, multi-site evaluations are less participatory in nature *vis-à-vis* intervention staff. In contrast to settings in which multi-center clinical trials are applied, multi-site evaluations address large-scale programs which because of their (complex) underlying strategies, implementation issues or other reasons are not amenable to controlled experimental impact evaluation designs. Possible variations in implementation among interventions sites, and variations in terms of available data require a different more flexible approach to data collection and analysis than in the case of the multi-center clinical trials. A common framework of questions and indicators is established to counter this variability, enabling data analysis across interventions in function of establishing generalizable findings (Straw and Herrell, 2002).

It should be noted that not only cluster evaluation but also multi-site evaluation are applicable to homogenous programs with little variation in terms of implementation and context among single interventions (i.e., cell 1 in Table 1).

Cell 2

Cells 2 and 4 cover a number of distinct synthesis approaches; multi-level evaluation studies that rely primarily (and often exclusively) on existing available evidence at project level as well as other sources of existing evidence. The systematic review is a term which is used to indicate a number of methodologies that deal with synthesizing lessons from existing evidence.[5] In general, one can define a systematic review as a synthesis of primary studies which contains an explicit statement of objectives and is conducted according to a transparent, systematic and replicable methodology (Greenhalgh et al., 2004). Typical features of a protocol underlying a systematic review are the following: defining the review questions, developing the protocol, searching for relevant bibliographic sources, defining and applying criteria for including and excluding documents, defining and applying criteria for assessing the methodological quality of the documents, extracting information, synthesizing the information into findings (see Oliver et al., 2005).

Meta-analysis is one of the best known systematic review methodologies and belongs in cell 2. A meta-analysis is a quantitative aggregation of effect scores established in individual studies. The synthesis is limited to a calculation of an overall effect score expressing the impact attributable to a specific intervention or a group of interventions. In order to arrive at such a calculation, meta-analysis involves a strict procedure to search for and select appropriate evidence on the impact of single interventions. The selection of evidence is based on an assessment of the methodology of the single intervention impact study. In this type of assessment usually a hierarchy of methods is applied in which randomly controlled trials rank highest and provide the most rigorous sources of evidence for meta-analysis. Meta-analysis differs from multi-center clinical trials in the sense that in the former case the evaluator has no control over the single intervention evaluations as such. As a result, despite the fact that homogeneity of implementation of similar interventions is a precondition for successful meta-analysis, inevitably meta-analysis is confronted with higher levels of variability in individual project implementation, context and evaluation methodology than in the case of multi-center clinical trials.

Meta-analysis is most frequently applied in professional fields as medicine, education, and (to a lesser extent) criminal justice and social work (Clarke, 2006). Knowledge repositories like the Campbell and Cochrane Collaboration rely heavily on meta-analysis as a rigorous tool for knowledge management on what works. Both from within these professional fields as well as from other fields substantial criticism has emerged. In part, this criticism reflects a resistance to the idea of a "gold standard" underlying the practice of meta-analysis.[6] This has been useful in the sense that is has helped to define the boundaries of applicability of meta-analysis and the idea that, given the huge variability in parameters characterizing evaluations, there is no such thing as a gold standard (see Clarke, 2006).

Partly as a response to the limitations in applicability of meta-analysis as a synthesis tool, more comprehensive methodologies of systematic review have been developed. An example is a systematic review of health behavior amongst young people in the United Kingdom involving both quantitative and qualitative synthesis (see Oliver et al., 2005). The case shows that meta-analytic work on evidence stemming from what the authors call "intervention studies" (evaluation studies on similar interventions) can be combined with qualitative systematic review of "non-intervention studies," mainly research on relevant topics related to the problems addressed by the intervention. Regarding the latter, similar to the quantitative part, a systematic procedure for evidence search, assessment and selection is applied. The difference lies mostly in the synthesis part which in the latter case is a qualitative analysis of major findings. The two types of review can subsequently be used for triangulation purposes, reinforcing the overall synthesis findings.

Cell 4

Systematic Reviews also feature prominently in cell 4. The more the implementation and the nature of interventions within a program or portfolio vary, the less feasible is a quantitative synthesis approach. In this sense the methodologies that belong in cell 4 are mainly based on qualitative synthesis approaches. In a way, the basic protocol for systematic review along the lines described earlier can also be used for exclusively qualitative data analysis and synthesis approaches.

Perrin (2006) presents a useful distinction between the synthesis of existing studies and the integrative review. The former is based on a series of studies from within an agency commissioning an evaluation, including among other things documents about the different single interventions that make up the program. There is not necessarily an explicit protocol underlying the review process and the exercise is basically one of summarizing findings from earlier studies and synthesizing these findings at the program level. A synthesis of existing studies appears to be similar to (or encompasses) what Pawson (2002a) labels the narrative review. Pawson contrasts narrative reviews with meta-analyses (which in a sense are also a syntheses of existing studies). While the latter in fact de-contextualizes interventions as it focuses on the aggregation of quantified effect scores, narrative review tends to over-contextualize interventions, providing ample attention to context and detail but (as argued by Pawson, 2002a) lacking a satisfying structure for the aggregation of lessons from individual projects, and as such inhibiting its capacity for generalizing findings across interventions. A narrative review is a descriptive account of intervention processes or results covering a series of interventions. Often, the evaluator relies on a common analytical framework which serves as a basis for a template that is used for data extraction from the individual studies. In the end, the main findings are summarized in a narrative account and/or tables and matrices representing key aspects of the interventions. The descriptive analysis of the review creates a kind of tension between the attention for details versus the comparison across interventions.

An integrative review is not restricted to the sources of evidence from within an agency. Instead, an evaluator resorts to a broad spectrum of available sources of evidence which can include basically everything: evaluative evidence from within the agency, published and unpublished research, grey literature on policy interventions, and so on. In order to establish whether certain findings are robust, the evaluator not only relies on the principle of assessing the quality of a study supporting the findings but also, and perhaps even more so, on triangulation among sources of evidence. As the potential universe of evidence to be taken into account is vast, an integrative review is usually based on a formal structure involving a set of rules for evidence selection and incorporation into the synthesis.

A relatively recent type of integrative review is the so-called realist synthesis (Pawson and Tilley, 1997; Pawson, 2002b). This particular approach in several ways is markedly distinct from other approaches. Realist synthesis emphasizes the embeddedness of interventions in complex social realities. How is this complexity addressed by the methodology?

The basic notion underlying realist syntheses is that interventions offer opportunities which may or may not be acted upon by different stakeholders involved in the intervention. Whether and how these stakeholders act upon the mechanism depends on a number of context-specific issues. Correspondingly, realist evaluators perceive of interventions as resources that trigger configurations of mechanisms, context and outcomes. An intervention mechanism is never implemented twice in the same manner. Apparently identical intervention formats (in terms of ideas, resources, planning) can produce very divergent outcomes given that the wider social and institutional context conditioning the implementation processes as well as processes of change is always different. The trinity context-mechanism-outcome constitutes the basis for the reconstruction of the assumptions underlying interventions and the subsequent testing and refinement of these assumptions. As a result, the "final" product of a realist synthesis is not a narrative nor is it a calculated aggregate effect size, but a theory about how an intervention or group of interventions works. This theory is never finished but always subject to further improvement. As such, the realist synthesis represents a specific approach to theory-based synthesis.

Current Practices of Generating Multi-Level Evaluative Evidence within the GEF Evaluation Office

In this section we present an inventory of the main evaluation modalities for generating (multi-level) evaluative evidence within the GEF Evaluation Office. In addition, we characterize these modalities using the methodological classification presented in the previous section. Table 2 provides an overview of the main types of evaluation implemented by the Evaluation Office. The table does not include individual project evaluations (appraisal, mid-term and ex post) which are the responsibility of the implementing agencies and serve as an input for the different evaluations managed by the Evaluation Office.

The most comprehensive and elaborate evaluation is the Overall Performance Study. An Overall Performance Study is implemented once every three or four years in preparation of a replenishment of the GEF. The main objective is to provide an overall assessment of the extent to which the GEF has reached its main objectives as stipulated in the various policy documents endorsed by the Council. Its approach is comprehensive in the sense that it covers all focal areas and looks at results as well as processes. In terms of methodology, the overall performance study is primarily a synthesis of existing studies. For example, the most recent

Table 2
Types of Evaluation Managed by the GEF Evaluation Office

Evaluation type	Main purpose	Underlying methodology
Overall performance studies	Accountability, learning in function of strategic changes	Synthesis of existing studies, Multi-site evaluation
Annual performance reports	Accountability, learning in function of improved implementation	Synthesis of existing studies
Program evaluations	Accountability, learning in function of strategic changes at focal area level, learning about project effectiveness	Synthesis of existing studies, Multi-site Evaluation
Thematic and process evaluations	Thematic learning	Multi-site evaluation, Synthesis of existing studies
Country portfolio evaluations	Accountability, learning in function of strategic changes at country level	Synthesis of existing studies, Multi-site evaluation
Impact evaluations	Accountability, learning about effectiveness and impact at the level of projects and (key) strategies	Multi-site evaluation

Overall Performance Study (OPS3) heavily relied on the findings of the different program evaluations (evaluations of focal area portfolios). A qualitative analysis based on the findings of all evaluative output of the GEF network constitutes the core component of the methodology. In addition, a stakeholder analysis of actors (individual and group interviews) at different levels of the GEF network, selected project field visits, and additional literature review are key components of the overall methodology.

The annual performance reports are syntheses of project evaluations, the latter falling under the responsibility of the Implementing Agencies (see Figure 1). Ex post terminal evaluations of individual projects are systematically processed by the Evaluation Office and analyzed on the basis of three dimensions: results on the basis of outcome and sustainability ratings; processes affecting results such as supervision by Implementing Agencies and co-financing arrangements; and the quality of project-level monitoring and evaluation, using the GEF M&E Policy as a benchmark. Furthermore, on process issues additional evaluative work may be done through interviews, surveys and analysis of agency documentation. Thus, the annual performance report is a meta-evaluation looking at both the content and quality of existing evaluative material at project level. As a synthesis of existing studies *pur sang,* it focuses on the quality of projects as well as the type and quality of information provided by the different projects. The latter allows for developing tailored recommendations on improving the alignment (type of information provided to the Evaluation Office) and aggregation (the way information is reported by the Implementing Agencies and processed by the Evaluation Office). In addition, the analysis allows for developing specific recommendations on project management by Implementing Agencies.[7] The depth of the analysis is inherently limited due to the substantial variability in interventions supported by the GEF and the type of data (i.e., terminal evaluations only) included in the exercise.

A third important type of evaluation concerns the program or focal area evaluations. The three main focal areas of Biodiversity, Climate Change and International Waters have been subject to an extensive program evaluation, so far once every replenishment period.[8] Similar to the approach of the Overall Performance Study, though with more attention to the specific issues and strategies within one focal area, a program evaluation first and foremost is based upon an extensive review of different types of available evidence within the GEF network on strategies within the focal areas, and different projects and themes addressed by projects. In this sense, the study is primarily a synthesis of existing studies. Several implementation-related, as well as results-related, topics are highlighted. Within one focal area the diversity in terms of types of intervention strategies and contexts remains substantial and a comprehensive in-depth assessment of all these issues goes beyond the (substantial) time and budget allocations for this type of evaluation. Consequently, usually a number of priority themes are selected for analysis. For example, among other things the Biodiversity Program Evaluation completed in 2004 highlighted the topics of management of protected areas and mainstreaming biodiversity (GEF, 2004). To supplement the desk study, the program evaluation relies on stakeholder interviews at different levels of the GEF network, additional literature review and selected field visits, the latter in line with priority issues highlighted by the evaluation. In preparation for the Fourth Overall Performance Study, the Evaluation Office will undertake evaluations of all six focal areas.

Thematic and process evaluations are somewhat different than the previously discussed evaluations in the sense that the focus is not on a certain policy level such as the focal area (i.e., as in Program Evaluations) or the entire GEF portfolio (i.e., as in the Overall Performance Studies) but on specific cross-cutting themes or topics. As a result, evaluations tend to be more focused on a particular theme which makes it easier for evaluators to do in-depth analyses of specific projects and synthesize lessons coherently into an overall evaluative picture. Examples of such more streamlined evaluations are the evaluation on projects that support countries in order to comply with the Cartagena Protocol on Biosafety (i.e., a thematic evaluation within the Biodiversity focal area) or the evaluation of the GEF Activity Cycle and Modalities (i.e., a process evaluation across focal areas and GEF partner agencies). A departure from this more narrow thematic focus is the recently completed Evaluation of Local Benefits in Global Environmental Projects. Here, the emphasis has been on the theme of local benefits to stakeholders,[9] looking at this theme from the perspective of major elements of the GEF portfolio. The Local Benefits Study is a typical example of a multi-site evaluation, with the main component being a substantial number of field visits (including stakeholder interviews) all over the world across different focal areas, involving different GEF agencies, focusing on the different mechanisms underlying GEF-supported interventions resulting (or not) in local benefits. Other thematic and process evaluations are primarily syntheses of existing studies, relying principally on desk study of existing documents and data sets, often complemented with stakeholder interviews, field visits and additional literature review.

An evaluation modality of increasing importance is the country portfolio evaluation. Recently, the GEF has adopted a new resource allocation system for the two main focal areas Biodiversity and Climate Change. To improve targeting from the perspective of Global Environmental Benefits, resources are now allocated according to country-level characteristics, more specifically: an index capturing the importance of the country in terms of (potential) global environmental benefits, and a country-level index capturing institutional performance. The new system has challenged recipient countries to prioritize the proposed interventions of the GEF. Encouraged by the GEF Council, the Evaluation Office has started a first Country Portfolio Evaluation in 2005 in Costa Rica. The scope of a country portfolio evaluation is broad, covering all GEF-supported projects in a country, thereby focusing primarily on: efficiency (e.g., efficiency of preparation and implementation of projects; of inter-institutional collaboration), relevance (e.g., relevance with respect to country priorities; the GEF's mandate) and data reporting on results (e.g., assessment of indicator frameworks). A country portfolio evaluation's methodology is essentially a synthesis of existing studies (i.e., the relevant documentation on all GEF-supported projects in a country), complemented by a multi-site framework of field visits to selected projects in combination with stakeholder interviews with representatives from different groups of stakeholders.

Finally, the Evaluation Office has recently embarked upon a first assessment of impact of GEF support, focusing on the Biodiversity focal area. During the planning phase discussions on the methodological framework revealed several critical methodological challenges (GEF, 2006b; Vaessen and Todd, 2007). As a result, it was decided that the impact evaluation should focus on one intervention strategy only, in this case support for protected areas, a key strategic area of intervention within the Biodiversity portfolio. In addition, a theory-based evaluation approach was selected as the best option to address some of the methodological challenges that were identified. Data availability in one particular case allowed for a complementary quasi-experimental impact evaluation of support for protected areas in Costa Rica (see next section). The main methodological component, the theory-based impact evaluation of protected area strategies, is basically a multi-site evaluation approach covering a relative homogenous set of interventions, in this case GEF support for three protected areas in Eastern Africa.

Multi-Level Evaluative Evidence within the GEF: Key Issues and Trends

In the previous section we made an inventory of the main evaluation modalities executed by the Evaluation Office, all of these being multi-level in terms of covering multiple (two or three) policy levels of intervention. Most evaluations are based on a methodology that combines some type of multi-site evaluation framework (a series of field visits to different projects, in different countries) with a synthesis of existing studies (a review of existing documentation on the project or program). Both methodological frameworks are flexible and allow for a broad ad hoc interpretation that serves the pragmatic needs of an evaluation office closely tied to the policy cycle of the facility it serves.

The above does not imply any normative statement about quality or rigor. This would be an empirical question requiring an in-depth analysis into the methodological setup of individual evaluation studies. What can be said, however, is that the flexible methodological designs inherent to multi-site evaluation and synthesis of existing studies do not have "built-in" "validity checks," methodological principles that help to ensure that threats to validity are adequately addressed.[10] Examples of such "validity checks" are protocols for assessing the quality of evidence affecting the decision to include evidence in an evaluation (as in systematic reviews) or statistical design principles to construct a framework for counterfactual analysis (as in multi-center clinical trials and meta-analysis).

Currently the Evaluation Office employs various ad hoc strategies to ensure the validity of data. For example, in the evaluative work for the Annual Performance Report, project evaluations are assessed systematically by staff of the Evaluation Office in order to ascertain whether the required quality is met to allow further use of data for the synthesis of results and processes affecting results. This is not yet systematically in use for other multi-site evaluations. In many of the evaluations at port-folio level, quantitative and qualitative data collected at project-level (that are deemed reliable) are further analyzed using specific methods. For example, on qualitative evidence, the Evaluation Office has started to employ Atlas.ti software to ensure statistically sound analysis of re-lationships between statements, leading to richer and better established synthetic findings. On quantitative data sets of project-level indicators, statistical tools are used for univariate descriptive analysis, bivariate descriptive analysis of associations and correlations between variables, and where possible statistical inference analysis.

Should the Evaluation Office then opt for this second category of methodologies with built-in validity checks? A tentative answer would be, yes, but only to some extent. In the remainder of this section we discuss why the scope for application of "more rigorous" methodologies is limited within the GEF Evaluation Office. At the same time we present some interesting applications of innovative methodologies that are applied by the GEF Evaluation Office to provide reliable answers to difficult evaluative questions.

In a recent paper Vaessen and Todd (2007) describe how different aspects regarding the nature of the evaluand within the framework of the GEF complicate evaluation efforts. First of all, the objective of the GEF is to support the generation of global environmental benefits. Yet, processes of environmental change are inherently complex (affected by multiple drivers), uncertain, and non-linear and therefore difficult to assess. Second, in practice the GEF implements several UN treaties such as the Convention on Biological Diversity and the United Nations

Framework Convention on Climate Change. Some of the principles and concepts within these treaties are difficult to operationalize, resulting in divergent interpretations across interventions. Third, there is a clear tendency that the GEF is shifting from site-specific interventions to broader support with the aim of advancing environmental concerns at (national and international) policy level. In the latter case, causal relationships between intervention and outcome are often more diffuse and difficult to capture. These issues cause difficulties for evaluation efforts at all levels within the GEF network, including the multi-level more aggregate exercises implemented by the GEF Evaluation Office. Although the above in principle does not preclude the application of any of the range of methods discussed earlier, it does explain to some extent why flexible methodological frameworks are preferred over more rigid frameworks based on the principle of measuring a narrow set of target variables.

A specific factor directly affecting the choice of methodology is the substantial diversity in terms of intervention activity and context characterizing the programs and portfolios evaluated by the Evaluation Office. This diversity is often further compounded by the lack of clarity about how individual intervention activities within a program feed into overarching program-level strategic goals (GEF, 2004; GEF, 2005). This basically disqualifies the use of methodologies under cell 1 and 2 of Table 1, for all evaluations that focus on levels such as focal areas or operational programs (a sub unit of focal areas) as unit of analysis. Neither is cluster evaluation (cell 3) particularly applicable. Multi-level evaluations focusing on focal areas, topics or processes often cover a large quantity of interventions involving a large number and variety of actors as well as geographical settings. More importantly, cluster evaluation would be more appropriate at the level of GEF Agencies as they are directly responsible for operational issues of GEF interventions, while there is quite some "distance" between the GEF Evaluation Office and operations.

Several evaluations have highlighted the data reporting deficit within the GEF network. Improved strategic guidance for example in the biodiversity focal area resulting in the definition of so-called strategic priorities (e.g., mainstreaming biodiversity in productive landscapes) and corresponding tracking tools are likely to have a positive effect on data availability and alignment between project-level and program-level data. However, this tendency is relatively recent, and many of the past and current ongoing projects lack standardized and adequate reporting systems on processes, outputs and outcome (GEF, 2004; GEF, 2005). In a similar guise, project evaluations often lack adequate data for further evaluative analysis (GEF, 2005). This situation substantially affects the prospects for any type of evaluative synthesis. More specifically, absence of (quasi-)experimental evidence at project (or intervention activity) level precludes the application of meta-analysis.

Limited budgets and time constraints require that the Evaluation Office deals with complexity, diversity and data constraints in a cost-efficient and flexible manner. To some extent evaluations have to sacrifice depth of analysis in favor of scope, trying to capture a comprehensive range of issues or range of geographical and contextual settings in which GEF interventions are implemented, to be able to present generalizable statements of merit and worth to the Council. In that sense, flexible multi-site evaluations and syntheses of existing studies present a logical choice as overarching approach.

Apart from these restrictions affecting the production of multi-level evaluative evidence a number of interesting alternative approaches to generating evaluative evidence have been piloted by the Evaluation Office. A first interesting development is a renewed interest in (quasi-) experimental evaluation. A recent example is a quasi-experimental evaluation of the impact of GEF support on protected areas in Costa Rica, within the framework of a recent impact evaluation managed by the GEF Evaluation Office. The analysis was based on existing time series data on deforestation in Costa Rica. The data allowed for a comparison of forestry cover over time between protected and non-protected areas. In addition, within the former group, protected areas that had received GEF support could be compared with protected areas without GEF support. As a result, the effect of GEF support on avoided deforestation could be approximated. This type of analysis requires that a number of threats to validity, such as the non-random establishment of protected areas, be explicitly addressed in the analysis. In the future, the Evaluation Office will continue to search for opportunistic possibilities for counterfactual analysis on the basis of existing data sets.

Strictly speaking, the type of analysis described above is not a multi-level evaluation. Nevertheless, there is a clear analogy with multi-center clinical trials described earlier. The latter are based on a series of randomized controlled trials. In a similar manner, one could conceive of a series of quasi-experimental evaluations on similar GEF interventions in different regions (e.g., support for protected areas in different countries). Consequently, a comparison of the findings would allow for robust conclusions on the impact of GEF support with a high degree of external validity.

A second methodological innovation concerns the application of theory-based evaluation. In the same impact evaluation mentioned in the previous paragraphs, the major exercise was constituted by a theory-based evaluation of GEF support to three protected areas in Eastern Africa. One of the defining elements of this evaluation was the approach to constructing a theory of change connecting project outputs to outcomes and impacts. Multiple sources of evidence were used to construct the theory: analysis of existing documents and data at project level, additional field studies, and stakeholder interviews. Structured reasoning on the basis of the projects' logframes, complemented by a broader analysis on outcomes translating into impacts (thereby taking into account the influence of external drivers of change), and finally an analysis on reduced threats affecting biodiversity targets, all contributed to an increased understanding of impact.

In several ways, the impact evaluation provides an interesting basis for further methodological innovation. First of all, the impact evaluation shows the roots of a theory of change on protected area management, a theory which in future evaluations could be reused as a basis for analysis and further refined on the basis of new findings. In addition, the theory-based analysis could be a basis of inspiration for future synthesis studies. In contrast to the impact evaluation which is mainly a multi-site evaluation with empirical analysis on the ground, much of the evaluative synthesis work comprises large portfolios and is based on existing documentation and data only. A theory-based synthesis starts out from a basic framework of causal linkages either focused on project process chains or project results chains (or a combination of both). These chains are continuously adjusted and enriched on the basis of further evidence. Several methodological steps can improve the quality of this type of evaluative analysis and subsequently the quality of the evidence that in the end is used to support evaluative conclusions. A focus on a specific type of intervention or process aspect that recurs throughout a portfolio is important. For example, a focus on capacity building activities, establishing trust funds or protected area management activities supported by the GEF is much more useful than a focus on operational programs or even projects (being administrative categories), the latter comprising a sometimes substantial diversity in intervention activities. This is in line with recommendations by several authors who argue for evaluation to be more focused on policy instruments as useful units of analysis for learning and accountability (Pawson, 2003; Bemelmans-Videc and Rist, 1998).

Another interesting methodological step not only in theory-based synthesis but in any type of synthesis would be the application of a clear and verifiable protocol which includes rules for evidence selection to be incorporated in the evaluation. These principles of systematic review are especially important in theory-based synthesis which ideally should benefit from the state of the art of (academic) knowledge on several key aspects relating to an intervention type or process aspect. Clear criteria for sifting through and selecting from the massive and diverse data bases of existing documentation are essential to ensure the incorporation of relevant high-quality evidence in synthesis studies, as such, in the end improving the quality of the evidence that is produced through the synthesis itself.

Concluding Remarks

In this chapter we have discussed different methodologies for generating multi-level evaluative evidence. The practice of multi-level evaluation within the GEF has revealed many challenges to the delivery of high-quality evidence, such as the availability of data, the coherence between project-level objectives and higher-level goals, and the planning and execution of cost-effective comprehensive evaluations with methodological designs that produce credible, reliable and generalizable findings. With respect to the latter, the Evaluation Office can rely on different strategies, for example: careful case and site selection in multi-site evaluations, systematic designs for data collection, diversification of and triangulation between different sources of evidence to support findings, or theory-based analysis to understand the causality underlying processes of change.

A particular aspect of the quality of evidence that requires further reflection is the tension between accountability and learning as principal purposes of the evaluation function. The Evaluation Office is committed to serve both purposes. In some cases, both aspects can be harmoniously united in a particular evaluation. For example, in a process evaluation, the relatively narrow focus on a particular process as unit of analysis (e.g., the GEF activity cycle) facilitates the coherent and systematic aggregation of lessons across interventions and contexts, thereby serving both learning as well as accountability purposes. In contrast, in program evaluations, accountability is rendered at the level of the program, which is a particularly difficult unit for aggregating lessons and further knowledge management, due to the enormous diversity in intervention activities, implementation processes, contexts and outcomes within a program. In terms of (multi-level) evidence, the type of evidence that is produced to allow for statements on merit and worth is often not the same type of evidence most useful for learning purposes. For example, the outcome ratings produced by project evaluations and summarized in the annual performance reports clearly satisfy accountability needs but are not very revealing about underlying dynamics of an intervention. The somewhat "softer" statements on impact produced by the theory-based impact evaluation on protected areas in Eastern Africa provide a promising example of reconciling both purposes in one evaluation.

Comparing the available multi-level evaluation methodologies with evaluative practice in the GEF Evaluation Office we have seen that the latter relies on the more flexible methodological frameworks rather than the more rigorous frameworks. This is due to several reasons such as the nature of the evaluand (e.g., the underlying uncertainty of environmental processes of change), the complexity of the evaluand (i.e., the variability in types of interventions within a portfolio), the (lack of) availability of adequate data at project level, but also practical considerations of resources, time and data availability and the specific demands within the organization for higher-level evidence on results and performance. Yet, even within these parameters there is scope for exploring new avenues to improve the quality of multi-level evaluative evidence.

In this chapter we have mentioned several options. Theory-based evaluation, both within the setting of multi-site evaluation as well as evaluative synthesis can be helpful in terms of providing a framework for reconstructing and connecting intervention processes to processes of change in different contexts. Currently, the GEF Evaluation Office is applying theory-based evaluation in multi-site evaluations such as impact evaluations of biodiversity projects and strategies. In general, evaluative work in the Evaluation Office increasingly relies on elements of realist evaluation as defined by Pawson and Tilley (1997): ensuring there is full recognition of the context and the diversity of the actions, focusing on the theory of underlying mechanisms supposed to bring lasting change and then bringing all available evidence to bear on both the context, the history of what happened and why it happened as such. With respect to evaluative syntheses, which in some way are part of most evaluations managed by the Evaluation Office, there is scope for further exploring theory-based synthesis approaches as well as applications of more systematic protocols for review and synthesis. The new developments in evaluative synthesis approaches, championed by the growing number of second order knowledge institutions and others (see Hansen and Rieper, this volume), produce interesting methodological insights for organizations like the GEF Evaluation Office. Vice versa, applications within organizations such as the GEF can leave their own particular mark on the development of evaluative synthesis methodologies, given the direct links to policy cycles and the specific and less than ideal circumstances (from a research perspective) in which these organizations operate.

Notes

1. The authors are respectively external consultant and director of the GEF Evaluation Office.
2. In practice, the two are often combined.
3. The so-called Enabling Activities, which support countries in their reporting to the multilateral environmental agreements, are evaluated by the Evaluation Office in the framework of a Country Portfolio Evaluation.
4. The classification includes methodologies that are useful for evaluating multi-level interventions that are either under implementation or have been completed and does not include ex ante multi-level methodologies such as particular multi-criteria analysis approaches or other types of modeling approaches (see Yang et al., 2004). In addition, it does not include multi-level information systems that are used for management purposes such as monitoring data on accounting, implementation progress or performance assessment.
5. Some authors use the term systematic review to refer to one specific type of methodology.
6. The critique against meta-analysis is mainly centered around the following points: lack of availability of (quasi-)experimental studies on single interventions as inputs to meta-analysis; inability of meta-analysis to include non (quasi-)experimental studies; the aggregation of results of sometimes different interventions; the simplification of outcomes; the reductionism of intervention contexts; technical and practical problems; ethical considerations (mostly related to randomization) (Clarke, 2006; Pawson, 2002a).
7. For example, the Annual Performance Report 2006 highlighted the differences in quality of supervision between the different Implementing Agencies, directly prompting a Council decision on improved project supervision by certain Agencies.
8. The smaller focal areas (in terms of budget) are subjected to more restricted evaluation exercises.
9. As opposed to global benefits (e.g., conservation of globally important biodiversity) which constitute the key focus of the GEF. One of the assumptions driving the evaluation was that global benefits are more effectively delivered when at the same time interventions produce tangible local benefits to different stakeholder groups.
10. Obviously, threats to the validity of findings of evaluations (see Cook and Campbell, 1979) are not something to be "solved" by methodology only. Ultimately, the (in)capable evaluator(s) determine(s) the quality of the overall findings.

References

Bemelmans-Videc, M.L. and R.C. Rist (eds.) (1998). *Carrots, sticks and sermons: Policy instruments and their evaluation.* New Brunswick: Transaction Publishers.

Clarke, A. (2006). "Evidence-based evaluation in different professional domains: Similarities, differences and challenges." In I.F. Shaw, J.C. Greene and M.M. Mark (eds.), *The SAGE Handbook of Evaluation,* London: Sage Publications.

Cook, T.D. and D.T. Campbell (1979). *Quasi-experimentation: Design and analysis for field settings.* Chicago: Rand McNally.

GEF (2004). *Biodiversity program study.* GEF Evaluation Office, Washington D.C.: Global Environment Facility.

GEF (2005). *OPS3: Progressing toward environmental results.* Third overall performance study, GEF Evaluation Office, Washington D.C.: Global Environment Facility.

GEF (2006a). *The GEF monitoring and evaluation policy.* GEF Evaluation Office, Washington D.C.: Global Environment Facility.

GEF (2006b). *GEF impact evaluations: Initiation and pilot phase—FY06.* Approach paper, GEF Evaluation Office, Washington D.C.: Global Environment Facility.

Greenhalgh, T., G. Robert, F. Macfarlane, P. Bate and O. Kyriakidou (2004). "Diffusion of Innovations in Service Organizations: Systematic Review and Recommendations." *The Milbank Quarterly* 82,1: 581-629.

IEG (2007). *The development potential of regional programs—An evaluation of World Bank support of multicountry operations.* Independent Evaluation Group, Washington D.C.: World Bank.

Kellogg Foundation (1991). *Information on cluster evaluation.* Battle Creek: Kellogg Foundation.

Kraemer, H.C. (2000). "Pitfalls of multisite randomized clinical trials of efficacy and effectiveness." *Schizophrenia Bulletin* 26: 533-541.

Lipsey, M.W. (2000). "Meta-analysis and the learning curve in evaluation practice." *American Journal of Evaluation* 21,2: 207-212.

Oliver, S., A. Harden, R. Rees, J. Shepherd, G. Brunton, J. Garcia and A. Oakley (2005). "An emerging framework for including different types of evidence in systematic reviews for public policy." *Evaluation* 11,4: 428-446.

Pawson, R. (2002a). "Evidence-based policy: In search of a method." *Evaluation* 8,2: 157-181.

Pawson, R. (2002b). "Evidence-based Policy: The promise of 'Realist Synthesis'." *Evaluation* 8,3: 340-358.

Pawson, R. (2003). "Nothing as practical as a good theory." *Evaluation* 9,4: 471-490.

Pawson, R. and N. Tilley (1997). *Realistic evaluation.* London: Sage Publications.

Perrin, B. (2006). "How evaluation can help make knowledge management real." in R. Rist, and N. Stame (eds.), *From studies to streams—Managing evaluative systems,* New Brunswick: Transaction Publishers.

Potter, Ph. (2005). "Facilitating transferable learning through cluster evaluation." *Evaluation* 11,2: 189-205.

Rist, R. and N. Stame (eds.). *From studies to streams—Managing evaluative systems.* New Brunswick: Transaction Publishers.

Straw, R.B. and J.M. Herrell (2002). "A framework for understanding and improving multisite evaluations." In J.M. Herrell and R.B. Straw (eds.), *Conducting multiple site evaluations in real-world settings,* New Directions for Evaluation 94, San Francisco: Jossey-Bass.

Turpin, R.S. and J.M. Sinacore (eds.) (1991). *Multisite evaluations.* New Directions for Evaluation 50, San Francisco: Jossey-Bass.

Vaessen, J. and D. Todd (2007). *Methodological challenges in impact evaluation: The case of the global environment facility*. IOB Discussion Paper 2007-01, Institute of Development Policy and Management, Antwerp: University of Antwerp.

White, H. (2003). "Using the MDGs to measuring donor agency performance." In R. Black and H. White (eds.), *Targeting development: Critical perspectives on the Millennium Development Goals*, London: Routledge.

Worthen, B.R. and C.C. Schmitz (1997). "Conceptual challenges confronting cluster evaluation." *Evaluation* 3,3: 300-319.

Yang, H., J. Shen, H. Cao and C. Warfield (2004). "Multilevel Evaluation Alignment: An Explication of a Four-Step Model." *American Journal of Evaluation* 25,4:493-507.

6

"User-Dependent" Knowledge as Evidence in Health Care

Laila Launsø and Olaf Rieper

Introduction

The Purpose of the Chapter

We want to make an argument for raising the awareness and priority given to the views and experience of those who are supposed to benefit from a given intervention. This is a concern noted by researchers in health care (Bury 2001; Frank 1997; Launsø and Gannik 2000; Paterson 2004; Paterson and Britten 2000, 2004). In particular, we examine the value of their knowledge of the outcomes of the interventions and assess the importance of their knowledge of the contextual settings in conditioning the outcomes of the interventions. The "users" are those to whom the intervention is addressed be they clients, patients, or whatever. The "outcomes" of the intervention are defined in this connection as any outcome that affects the users, be it according to stated goals and objectives or not, or be it positive, negative or no outcome.

The purpose of the chapter is to address the low position of users' knowledge in evidence-based policy and practice in general and in health care especially. We argue that users' knowledge should have a more privileged position in the hierarchy of evidence in situations where excluding the users' knowledge narrows down and restricts the potential for improving the service. In the service quality movement, for example, the users' knowledge and assessments are regarded as a valuable asset for improving the services (Kirkpatrick and Lucio, 1995).

This chapter is based on research in the treatment of chronic diseases involving conventional as well as unconventional treatments. The point, however, is not to privilege one set of treatments over the other, but to bring out the importance of the users' knowledge. Most, but not all, chronic diseases are complicated entities for which at a given point in time no known cure may be available, and often the causes of the disease are unknown. The chapter is built on research from health care, but the general characteristics of the issue are to be found in social work, education and many other services, see below.

The Users' Knowledge as Data

Users may be seen to be the prime observers of the outcomes of an intervention they receive, because it is in or around the individual the outcomes take place. Students, for example, are supposed to learn that they are to become better at reading, calculation, reflection, and cultural understanding. These are all competencies that are installed in and perceived by the individual. Patients are supposed to be cured or healed or at least to be getting a better life because of treatment, preventive and/or health promotion interventions. These outcomes are, for example, less pain, better coping with chronic diseases, or healthier lifestyles. The user normally has the ability to make self-observations of these matters and the user is in the exceptional position to experience the outcome shortly as well as long after the intervention has ended. A study on patients' choice of asthma and allergy treatments showed that users were applying a method of trial and error to test out the intervention and its possible outcomes compared to other interventions and compared to not using an intervention at all (Jørgensen and Launsø, 2005).

Despite the strength of users' knowledge of the outcome of an intervention these kinds of data have been put low down in the hierarchy of evidence. The hierarchy of evidence for selecting primary studies for systematic reviews typically looks as follows:

- Randomized controlled trials (with concealed allocation)
- Quasi-experimental studies (using matching)
- Before-and-after comparison
- Cross-sectional, random sample studies

- Process evaluation, formative studies and action research
- Qualitative case studies and ethnographic research
- Descriptive guides and examples of good practice
- Professional and expert opinion
- User opinion

We want to make the plea that the very low position of the users' knowledge of outcomes (user opinion) makes it difficult to include the users' voice in developing and adjusting interventions. The hierarchy of evidence may well be justified when one is dealing with technical interventions that are supposed to work directly and independent of the users' consciousness and activities. However, for many other kinds of interventions working indirectly and depending on the users' consciousness and self-activities and with complicated causal processes embedded in social system dynamics, the users' knowledge is valuable and in some situations the only source of data to rely upon. In these situations the effectiveness and benefit of the intervention are mediated by the behavior, actions and values of users themselves. Whilst RCT designed investigations clearly generate evidence that is scientifically valid they also unjustifiably leave out the kinds of user-dependent data that might lead to more satisfactory care for individual patients.

We are aware that RCT designs do not per se exclude information from the users, but most RCT designs that include users' information are carried out with a predefined questionnaire that leaves almost no space for user-defined categories and perceptions.

The Structure of the Chapter

The chapter first inserts the users' knowledge in a broader context in developing the quality of public service. Second, we mention the roots of the hierarchy of evidence from medicine as a contributing factor to the low status of users' knowledge. Third, we argue that even within health care, users' knowledge in relation to chronic diseases should be regarded as an important source of data. As mentioned above, our main focus is on health policy and practice, but similar arguments are supposed to be valid for other areas as well, such as social work and education. Fourth, we present from a research project a typical case of an individual who has attended various therapists. The case shows her experience with, and assessment of, the therapies and the consequences for her situation, her coping behavior and her understanding of the disease. Fifth, we present findings from a larger formative research-based evaluation conducted with the aim of improving the intervention for people with multiple sclerosis. This evaluation illustrates that the users' knowledge becomes valuable in understanding how team-based care is taken on board by the users. Both cases illustrate the importance of the context in which the intervention takes place and that information of this contextual setting can be more fully obtained through the knowledge of the end user.

The User-Orientation in the General Context of the
Service Quality Movement

Since the many and varied "users in focus" initiatives within public welfare services in the 1980s, users' satisfaction with services has been mapped through surveys conducted across wide areas of public (and private) service provision often related to the service quality movement of the public sector reforms in various countries. One can distinguish between at least three approaches to public service quality (Naschold, 1995).

First, a scientific/expert-led approach, in which quality standards are set by experts who ensure that the standards are made independently of the user/customer, albeit with a view to the "best interests" of the service user as determined by the expert. Various national and technical standards reflect this approach, such as standards for medical treatment, reference systems and building codes. This approach has a long tradition within the more technical and professional public services. The meta-analysis tradition within the evidence movement reflects this approach.

Second, a managerial (excellence) approach in which the aim is "getting closer to the service users" to deliver what they value most. Tools and models such as customer satisfaction surveys, quality of life questionnaires, client and staff consultation and total quality management standards are being used by an increasing number of public sector organizations in many countries. This is a powerful approach within the new public management (Pollitt and Bouckaert, 1995).

Third, a consumerist approach with the aim of empowering and engaging the users/customers through, for example, consumer rights (the UK Citizens' Charter championed by the Conservative government in the 1990s) and consumers' choice leading to competition among public service providers, but also through the development of a more active role for the user in service delivery (Kirkpatrick and Lucio, 1995, pp. 278-279). This last approach has two tracks: one is minimalist, guaranteeing a minimum level of quality, while another is a capacity-building one, in which the participation of the user in the service design, production and evaluation becomes essential, for example, in the care for elderly and in the health sector in general, where the credo is that elderly patients/users should be "empowered" to become more and more self-activating. In fact, the very use of the concepts "user" and "customer" indicates a shift of roles between the citizen receiving services and public organizations as providers of those services. In health care the sign of this shift is the focus on the individual as his or her own "disease manager." This shift also underlines the importance of users' knowledge about their disease and self-activity as well as users' assessment of treatments. This chapter is based on the latter track.

Within program evaluation "user or client models of evaluation" have been a recognized way of collecting assessments of services among those who are supposed to benefit from the services. The so-called "user-evaluations" have often been conducted as customer satisfaction surveys and have been criticized for being too simplistic and not able to catch users' assessment of the outcomes of the services in a valid way (Mayne and Rieper, 2003). Qualitative methodologies have been used as well, and they have also been criticized, e.g. for being based on the assumption that the service users are able to make a "correct" judgment of the outcomes of the intervention.

The main interest on the part of the user in evaluating public service quality should be to make informed decisions about which services to use, where such choices are assisted by information on actual levels of service provided, and if user fees are involved, to know if the service provides value for money. Another more direct strategy on the part of service users is to influence service quality through organized efforts by consumer associations and social movements. These associations and movements often have a strong interest in the design of evaluations and the use of evaluation findings to reshape the concept of service quality. This offers new possibilities for evaluations in shaping as well as testing out the ideas of consumer associations and movements in controversial fields of services (Launsø, 1996; Touraine 1981; Melucci, 1989; Eyerman and Jamison, 1991). In fact, one of the cases in this chapter is based on a project run by a patients' association (Haahr and Launsø, 2006; Launsø and Haahr, 2007).

"User-Dependent" Data in Health Care

The evidence hierarchy originates from the health field with a tradition for "objective" measurement of outcomes, with a focus on data that are observed by a third person independent of the knowledge of the users of a specific intervention. Users' knowledge is traditionally labeled "subjective," that is not trustworthy. An alternative view to this binary concepts "objective" and "subjective" with the concepts *"user-independent"* and *"user-dependent"* data. We consider these two last concepts as less normative than the first two concepts and add symmetry of importance to the concepts.

Looking into health sciences, the intention of evidence-based health sciences (EBHS) is to provide clinicians with a resource aiming at identifying and using the "right" intervention; that is the intervention that generates the optimal outcomes for the patient (see Foss Hansen and Rieper in this volume). Within the conventional health-care system the "right" intervention has increasingly been defined by the experts as a technical intervention. Accordingly, the evidence-based health science has gradually reduced to RCT-based meta-analyses relying on expert-defined interventions and outcomes. The users' knowledge is in this approach almost absent and the role of the physician in articulating users' knowledge has been marginalized. The technical intervention and "user-independent" data are of course relevant in situations where the expected outcomes are not dependent on an interaction between the intervention and the social system (the human being and her/his living context) involved and the users' consciousness, observations and experience. In developing this kind of EBHS the risk is that some ways of knowing are eliminated and especially in this context the knowledge of the users is eliminated. Why is this development a problem?

It is a problem because the intention to find the "right" intervention in producing the optimal outcomes for the user is based on too "thin" a stream of knowledge, which tends to preclude

1. "the knowing users'" own observations, experience, reflection and assessment of outcomes related to his living contexts,
2. the broader configuration of factors (besides the actual intervention) that affect the wider outcomes in the users' everyday living,
3. a broader concept of knowledge than expert-defined knowledge.

By eliminating the "knowing user" as the focus of the intervention—in producing and testing interventions, we assume that the scope of potential interventions and outcomes will be limited. This limitation does not support the aim of the health-care system, or other welfare services for that matter. The elimination of the knowing user is challenged by—among other things—the emerging global phenomenon: "the user-driven innovation" putting the "user-dependent" knowledge on the agenda.

Within health care the development of EBHS can be questioned as producing the whole "truth." One of the reasons is that the modern user of health care is using and experiencing a pluralistic health care including services within the public sector and within the private sector providing very different disease understandings, different diagnoses, different interventions and different outcomes. More and more users develop a reflective choice of interventions based on previous experience, assessments and information available on the Internet concerning other users' experience, assessments and knowledge (Sandaunet, 2008).

It can be argued that users' observations and judgments are especially important in cases with the following characteristics.

- First, the problem for the user (that the intervention addresses) is difficult to boil down to a very specific and well defined problem (in health care: diagnosis).
- Second, interventions do not produce outcomes independent of the social system (the living contexts) in which the intervention is "planted." Quite the opposite, the individual's self-activities and processes within the social system with its social norms and culture, are heavily influential on the obtainments of any positive outcomes. Thus, in these cases interventions seldom produce outcomes directly through linear causal processes. The outcomes are produced through learning processes and reflections that take place in the individual and in the social system. This is obvious in social work and in education, and should also be obvious within major parts of the health-care sector.

The Sussi Case

With these characteristics as outset we present a case from the health-care sector. The case illustrates a delicate interplay between interventions and outcomes and consequences for the demand on different forms of knowledge (Brendstrup and Launsø, 1997).

The Sussi case

Sussi is 36 years old and works as a kindergarten teacher. After having undergone a conical section she began having mild headaches. They later increased in severity. The operation developed to be a traumatic experience for Sussi. At a screening she had been informed that there were cellular changes and after the operation she was told that not everything had been removed. Having been discharged she had four successive hemorrhages at home. Returning to the hospital, she was treated and sent home immediately. Sussi received no information from the hospital and was afraid that it was very serious. She had a polypus removed and tested for cancer. She received no further information. Sussi found that it was hard to ask the doctors about the tests, and when she finally found herself face-to-face with a doctor she "almost felt like an idiot because she hadn't handled it herself." She was examined and got the message that "there is nothing left." During the four years following this course of events the migraine and the gastric acid imbalance increased in severity.

At an earlier stage Sussi had consulted her doctor about her headaches. "The doctor felt the back of my head and told me that the muscles were tense, and then he sent me to a physiotherapist." Sussi reported that she had seen the physiotherapist several times (she received 10-15 treatments). "It relieves for a moment, but when the treatment stops, it doesn't help any longer." Sussi took aspirin about every second week to reduce the most severe pain. At the end she felt that she had no life at home and she thought "what do I do?" "It was a magic day when I sat down and said to myself "enough is enough. I don't want to live like this anymore" and I reached for the phone book and searched for reflexologists in the area."

Together Sussi and the reflexologist tried to identify a possible cause of the headaches, and Sussi described what she had been through. "I went to see the reflexologist three times, and after that I noticed a fundamental change … as if some things in my body had been loosened.… I felt the pressure on my left shoulder, at the back of my head and behind my left eye. It was as if this pressure had been treated … like if a piece of clay is stuck inside and creating a pressure and then you remove it bit by bit—and it loosens. That was what happened after the fourth treatment. I guess the feeling was that I was cleansed from beneath and then it slowly moved upwards.…"

Sussi reacted very strongly to the treatments. She sweated a lot, she had numerous and large bowel movements and her urine had a strong smell. But at the same time she had the feeling of walking on cotton after a treatment. "When the headaches started to disappear I told myself to calm down, because I expected them to return quickly—I actually did. But I just kept getting better … it was like being reborn.… I had been spending a lot of time feeling unwell—time I should have spent with my family. It was really like being back in business. I actually think that it was my own mental process that made a change."

The gastric acid imbalance didn't disappear entirely until the headaches had disappeared. Sussi learned to press a key point in her ear, if she got a bloated stomach. She could feel that the key point was very tender in the beginning, but as she worked on it, the point got less tender. In addition, the reflexologist has helped Sussi to get rid of her heartburns—a problem that had previously required a lot of medicine. Sussi has been suffering from colds and throat inflammations since childhood, and she believes that the reflexology has had a positive outcome on these problems. It is now a long time since she used penicillin.

Sussi strongly reduced her intake of coffee, but only after having received five to seven treatments. She says that her body tells her what is good and what is bad for her.

After having engaged in the therapeutic process with the reflexologist, Sussi comprehends her headaches in another way. Whereas she earlier conceived the headaches as something she wanted to rip out, she now has a holistic understanding of her body. She says that if she should experience new symptoms in the future, her first reaction will be to search for possible explanations for these symptoms.

Initially, Sussi had the understanding that the reflexology specifically aimed at treating the headaches. But she has reconsidered this view; she has experienced that different symptoms can be related and that reflexology can have an outcome on the entire body.

The case illustrates a number of significant elements of information which can only be derived from the user's knowledge. First, the problem of the user, her symptom (the headache) was transformed from a specific entity to a broader contextualized phenomenon. Second, the first intervention that was applied was limited and did not take the user's perspective and knowledge into account. The subsequent intervention (reflexology) appears to have been more suitable and included social interaction between the user and the therapist. This intervention focused on the wider context of body-mind and activated the user's awareness and preventive behavior. Thus, self-regulating activities were set in motion, representing complex causal processes that led to positive outcomes (reducing the original problem of the user). In fact, it is difficult to imagine that self-regulating activities could be investigated without the elaborated "voice of the user."

Assumptions on Causality

Many users and therapists as well as researchers are socialized to think that it is the technical intervention in itself that determines the outcomes. Evidence-based medicine is normally based on this assumption (Gannik and Launsø, 2000). In contrast, the Sussi case shows that the communication with the therapist, Sussi's change in understanding her symptoms, and her own efforts are ascribed significance for the outcomes obtained. It is known, that the results of intervention are to be considered in connection with the communication between the therapist and the patient and the life situation (contexts) of the patients ((Kaptchuk et al., 2008; Gannik, 2005; Ostenfeld-Rosenthal, 2007; Antonowsky 1988). Also within psychotherapy it is well known that the specific form of therapy has less effect than the contextual and communicative aspects have (Wampold, 2001).

Figure 1 and 2 illustrate two different assumptions about what factors determine cure and relief (Launsø et al., 2007). Figure 1 contains the assumptions behind evidence-based medicine concerning the weight ascribed to each of the three scopes. Scope 1 (the influence of the intervention) is considered to have the most important direct influence, whereas scope 2 and 3 are ascribed much less weight. The size of the scopes is meant to be relative. Scope 2 and 3 are in evidence-based health sciences labeled placebo. Placebo is defined as a treatment without a specific effect. The issue is what constitutes a non-specific effect and this issue is much discussed in the literature (Barfod, 2005; Hrobjartsson and Gøtzsche, 2001; Kaptchuk, 2002). In Figure 2 it is scope 3 (the patient's own efforts) that is ascribed to have most influence on the cure and relief of the patients, whereas the intervention in itself is ascribed much less weight.

The three scopes are supposed to be in a complicated interaction with each other, for example, the outcome of the intervention is influenced by the communication between the therapist and the patient and the patient's own effort. Following this it might be difficult to assign the outcome to one of the three scopes separately.

Figure 1
Assumption behind Evidence-based Medicine

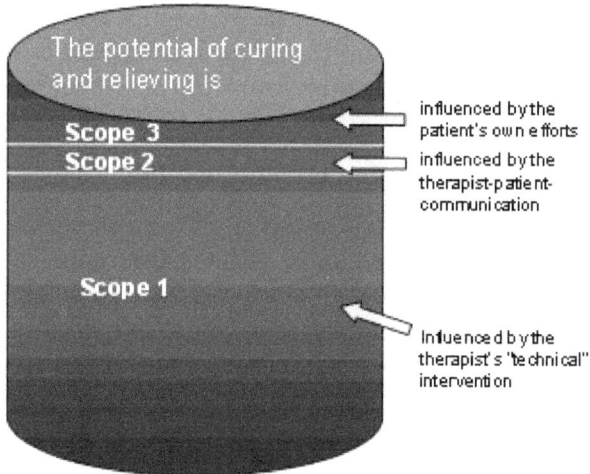

Figure 2
Assumption behind Clinical Practice

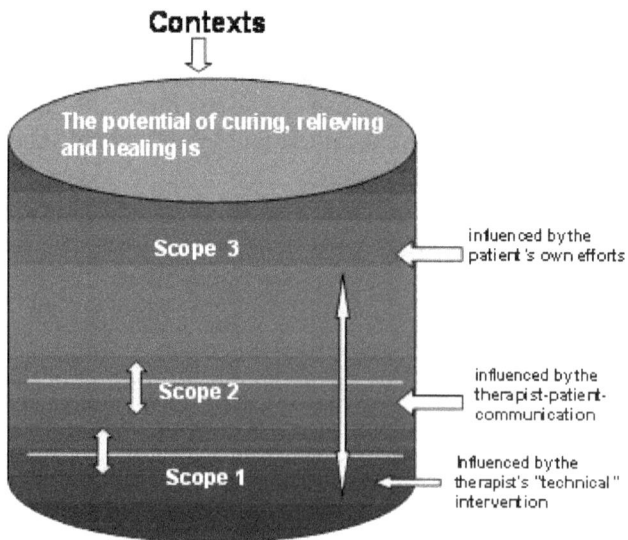

The assumptions of causality are different in the two figures. Figure 1 is based on an assumption of a linear and direct relation between the intervention and the outcome. Figure 2 is based on a more configurative or generative assumption of causality (Scocozza, 2000; Pawson, 2006).

In the following section we present a research project which illustrates various mechanisms and contexts and the significant role of the knowledge of the users.

Team-Based Treatments of Multiple Sclerosis— and Evidence-Based Knowledge

In the following section we provide a short description of the disease Multiple Sclerosis (MS). We then describe a research-based evaluation project of team-based treatment developed for People with MS (PwMS) and focus on the strengths and limitations of evidence-based knowledge and "user-dependent" knowledge in developing interventions oriented towards the users' daily living with a chronic disease characterized by a complex and broad spectrum of different symptoms.

The Disease Multiple Sclerosis (MS)

Multiple Sclerosis (MS) is a chronic and progressive neurological disorder that can lead to severe disability. The etiology of the disease is unknown (Li et al., 2004; Munger et al., 2004; Riise et al., 2003; Sørensen et al., 2004). In Europe, about 400,000 persons suffer from MS (EMSP 2004). MS cannot be cured by any known conventional treatment, but, in some cases, the progression of the disease can be slowed down medically, and a number of secondary symptoms can be treated medically (Sørensen et al., 2004). However, the symptomatic treatments are generally characterized by having only partial outcomes on the symptoms, and by having a number of side effects (Sørensen et al., 2004).

MS symptoms are attended to by a range of health professionals and treatment modalities. These treatment modalities are generally not coordinated in a plan of treatment and rehabilitation. A report from EMSP1 (2004) underpins the importance of interdisciplinary treatment approaches to be applied in MS. The traditional research activities in this field are characterized by testing single interventions.

Conventional and Unconventional Treatment

A survey of systematic reviews concerning conventional intervention has been conducted based on the Cochrane library covering the period 1998-2005.[2] Fourteen reviews were found. The results of these reviews indicated that the "evidence" for most of the conventional treatment of PwMS seems to be insufficient. Moreover, in the cases where the reviews detect outcomes there is no information on what kind of patients that might have low/high benefit from the treatment.

An increasing use of unconventional treatment is documented among PwMS (Marrie et al., 2003; The Danish Multiple Sclerosis Society, 2002, Storr, 2002; Haahr and Launsø, 2006). Using a method of trial and error, many PwMS combine conventional and unconventional treatment, without any counseling or advice offered in the health-care system due to—among other things—the lack of knowledge based on research. From PwMS there has been an increasing demand upon The Danish Multiple Sclerosis Society (a patient organization) to initiate a research project dealing with bridge building[3] and team-based treatment. The core question raised has been: Is it possible to improve treatment outcomes for PwMS by developing an operational model for bridge building between conventional and un¬conventional practitioners, and by developing a team-based treatment approach in practice?

Team-based treatment aims to restore and prevent physical, emotional, mental, energetic, spiritual and social unbalances based on the body's innate ability to heal (Launsø, 1996; Mulkins and Verhoef, 2004). This definition reflects an understanding of causality which is "situated" in the regenerative capacity of body processes and not in the intervention as such. Accordingly, treatment outcomes depend on an interaction between regenerative body processes, social contexts in which a person is embedded, and the treatment intervention. The point is that the treatment intervention itself is not assumed to be the determining cause of treatment outcomes. Further, team-based treatment is a clinical practice that ascribes importance to the relationship between practitioner and patient as well as the patients' self-activities.

Some patient associations and health-care providers demand new types of health care that encompass different concepts of health and disease, different diagnostic systems, different treatment methods and outcomes, and that combine conventional as well as unconventional practitioners (Launsø, 1989, 1996; Bell et al., 2002; Scherwitz et al., 2003; Mann et al., 2004; Mulkins et al., 2003, 2005; Launsø and Haahr, 2007).

Searching the literature published within the last five years did not yield any publications describing a research-based evaluation of team-based treatment of chronic disease, in which conventional and unconventional practitioners develop and offer treatments.[4] In the United States and Canada, team-based treatment approaches have been assessed by patients, practitioners and stakeholders (Mulkins et al., 2003, 2005; Scherwitz et al., 2003; Hollenberg, 2006; Coulter et al., 2007). However, in the evaluations, the outcomes were not related to the treatment combinations used by the patients. Further, these teams were characterized by a number of practitioners (conventional and unconventional) operating in parallel and referring to each other.

A Case of a Team-Based Treatment Research Project

In the light of these issues, the Danish Multiple Sclerosis Society initiated a bridge-building and team-based treatment project to take place from 2004-2010 at a specialized MS hospital.[5] In this project, a team of five conventional and five unconventional practitioners has been set up to work together in developing and offering individualized treatments to 200 PwMS at the MS hospital. The conventional practitioners (all staff members of the MS hospital) comprise one medical doctor (neurologist), one occupational therapist, one physical therapist, one psychologist, and one nurse. The unconventional practitioners included in the team are one acupuncturist, one nutritional therapist, one classical homeopath, one cranio-sacral therapist, and one reflexologist.

Patients were included from the population of referred patients at the MS hospital. The treatment of the participating PwMS takes place at the hospital during an initial hospitalization period of five weeks. Subsequent treatments take place at the hospital on an outpatient basis or in the practitioners' clinics, depending on the geographical residence of the participants. Patients are included on an ongoing basis from May 2005 to September 2007. During a period of six to 12 months the patient receives the usually-offered conventional treatments, combined with approximately 15 unconventional treatments. Each patient is included in the research project for 18 months. All the patients are followed from before start of treatment and up till 18 months after start of treatment.

The research design used is formative evaluation using quantitative and qualitative research methods. Both practitioners' and users' experience, reflections, knowledge development and learning processes are in the core research questions. Team meetings take place currently and four times a year there are researcher-practitioner seminars. At these seminars the different treatment models and the development of team-based treatment approaches are discussed and assessed in relation to treatment course of individual PwMS (Launsø and Haahr, 2007). In the collaboration with the practitioners the conceptual focus is on the description and discussion of the intervention, mechanisms, contexts, expected and obtained outcomes.

What Characterizes the PwMS's Experience with the Team-Based Treatment Approach?

Users' experience with outcomes of the team-based treatment approaches in the case of MS is collected through qualitative interviews and re-interviews with a sample of the users and through five rounds of questionnaires filled out by users throughout a period of 18 months. Qualitative interviews have given an insight into the facilitating and prohibiting conditions in obtaining positive or negative outcomes of the team-based interventions.

Interviews with the users have revealed that they perceive various kinds of mechanisms triggered by the intervention as well as contexts of importance for obtaining positive outcomes. The kind of mechanisms and context mentioned by the users is the following:

User-related mechanisms:

- *Mental mechanisms* including understanding of how an intervention is supposed to work, acceptance and knowledge of the significance of own efforts, opening up for repressed emotions, formulating own needs.
- *Behavioral mechanisms* including implementing own preventive activities, training, sleeping habits, diet.
- *Social involvement and activity.*

Treatment-related mechanisms:

- *The therapist's communication.* For example, the therapist uses time for communication, explains possible treatment options, is responsive to the needs of the user and respects the user. The therapist explains carefully the expected outcome, reactions on and side effects of the treatment given and how the user might cope with side effects of reactions.
- *Treatment mechanism* including user perceptions of changes in the body that last for a long time after the treatment, user perceptions of a higher mental energy level, and improved self-activity

Contextual aspects:

- *Related to user* including state of nutrition, motivation and readiness to participate actively in the treatment course, general ability to cope with problems in daily life, previous experience with treatments.
- *Related to social network* for example friends, loved ones.
- *Related to treatments for example* coordination and communication among the various therapists, mutual knowledge among the therapists about the other therapists' treatment models and approaches.

What Kind of Knowledge is needed from a User-Perspective?

Users judging the outcomes of the team-based treatment to be positive, stress the importance of their own rehabilitation activities. They point to the significance of support by the therapists and that the therapist should take into account the resources and the situation of the user in order to help developing competence as MS manager. Those users are thinking, so to speak, within the assumptions in figure 2. However, other users expect the therapists to manage their disease. These users experience training and rehabilitation as unpleasant interventions in their daily living, and think more in line of figure 1.

Users believe that users with chronic diseases are very different and that these differences should be made clear and respected by the therapists. They believe research and therapy should take as a point of departure the users' knowledge on the mechanisms and contexts that facilitates a positive outcome of treatments. The RCT base in evidence-based practice excludes research on the rich user-dependent knowledge on mechanisms and contexts of relevance for choosing the treatments with positive outcomes. The MS case and the Sussi case indicate that not taking the users' own knowledge into account leaves us with a poor research strategy and with research results that lack external validity.

Conclusion and Discussion

Service users' perspectives are an integral aspect of the service quality movement, but seem to be marginalized in the evidence movement. Also in the service quality movement, users' knowledge is seldom taken seriously, but information from the users is typically limited to user satisfaction survey or other instruments with predefined categories.

We have tried to argue for an upgrading of the users' voice in the evidence movement where users' experience has a low ranking in the hierarchy of evidence.

The chapter deals with health-care practice because in this area the argument for evidence-based practice has echoed most loudly, though opposing views have also been aired. Health care is therefore a critical test case. We believe, however, that the need for user-dependent knowledge can also be extended to other service areas.

Evidence-based health care or evidence-based medicine usually stresses that evidence should be based on a meta-analysis of studies with RCT designs. Mostly, meta-analysis also applies quantitative methods of data collection and analysis and these methods are considered more valid (objective) than other methods.

However, examples from research and evaluation of interventions to people with chronic diseases show that the voice of the user (patient) contains a rich source of information on outcomes, mechanisms and context of the interventions received. This is a form of knowledge that is not accessible through pre-structured and quantified measurement, but has to be based on qualitative interviews with or diaries from the service users.

This knowledge provides another kind of evidence than is usually applied in evidence-based medicine (and other evidence-based practices) and is a valuable source to understand what works in the interventions for people with complex problems, why and in which contexts.

Notes

1. The European Multiple Sclerosis Platform (EMSP) is an association of national MS societies.
2. The survey was done as part of a study on the evidence movement financed by the Danish Social Science Research Council (internal document: Notat vedr. beskrivelse af Cochrane-reviews om sklerose, 5/7 2006).
3. "Bridge building" refers to interdisciplinary cooperation at institutional level (public as well as private) between practitioners. More than just cooperation, bridge building requires the representation of different treatment models.
4. The search was carried out in the databases PubMed MEDLINE, EMBASE and The Cochrane Library, using the key words: Integrated/Integrative/Unconventional/ complementary/supplementary + medicine/treatment/care.
5. The project takes place in "The MS Centre at Haslev," one of two specialized MS hospital facilities in Denmark. At the other centre "The MS Centre at Ry," a group of PwMS is defined as a comparison group. Geographically, the centers are placed in the eastern and western part of Denmark, respectively.

References

Antonowsky, A. (1988). *Unraveling the mystery of health. How people manage stress and stay well.* San Francisco: Jossey-Bass.

Barfod, T.S. (2005). "Placebos in medicine: Placebo use is well known, placebo effect is not." *BMJ* 330 (7481):45.

Bell, I.R., C. Oper, G.E.R. Schartz, K.L. Grant, T.W. Gaudet, D. Rychener, V. Maizes and A. Weil (2002). "Integrative medicine and systemic outcomes research." *Archives of Internal Medicine* 162 Jan. 28 (reprinted).

Brendstrup, E. and L. Launsø (1997). *Headache and reflexological treatment.* Copenhagen: The Council Concerning Alternative Treatment, The National Board of Health.

Bury M. (2001). "Illness narratives: fact or fiction." *Sociology of Health and Illness* 23,3: 263-285.

Coulter, I.D., M.A. Ellison, L. Hilton, H. Rhodes and C. Ryan (2007). *Hospital-based integrative medicine. A case study of the barriers and factors facilitating the creation of a center.* CA: RAND Health.

EMSP (European Multiple Sclerosis Platform) (2004). *Recommendations on reha-bilitation services for persons with Multiple Sclerosis in Europe.* Multiple Sclerosis International Federation.

Eyerman, R. and A. Jamison (1991). *Social movements—a cognitive approach.* Cambridge: Polity Press.

Frank A. 1997. "Illness as a moral occasion; restoring agency to people." *Health* 1,2:131-148.

Gannik, D.E. (2005). *Social sygdomsteori—et situationelt perspektiv.* [Social disease theory—a situational perspective]. Copenhagen: Forlaget Samfundslitteratur.

Gannik, D.E. and L. Launsø (eds.) (2000). *Disease, knowledge and society.* Copenhagen: Forlaget Samfundslitteratur.

Haahr, N. and L. Launsø (2006). "Preliminary initiatives in a bridge building project between conventional and alternative practitioners in Denmark." *Forschende Komplementärmedizin und Klassische Naturheilkunde* 13:307-312.

Haahr, N. and L. Launsø (2007). "Bridge building and integrative treatment of persons with Multiple Sclerosis. Research-based evaluation of a team building process." *Journal of Complementary and Integrative Medicine* (JCIM) (Accepted for publication).

Hollenberg, D. (2006). "Uncharted ground: Patterns of professional interaction among complementary/alternative and biomedical practitioners in integrative health-care settings." *Social Science and Medicine* 62,3: 731-744.

Hrobjartsson, A. and P.C. Gøtzsche (2001). "Is the placebo powerless? An analysis of clinical trials comparing placebo with no treatment." *New England Journal of Medicine* 344,21:1594-1602.

Jørgensen, V. and L. Launsø (2005). "Patients' choice of asthma and allergy treatments." Journal of Alternative and Complementary Medicine 11,3: 529-534.

Kaptchuk, T.J. (2002). "The placebo effect in alternative medicine: Can the performance of a healing ritual have clinical significance?" *Annals of Internal Medicine* 136: 817-825.

Kaptchuk, T.J.; J.M. Kelly, L.A. Conboy, R.B. Davis, C.E. Kerr, E.E. Jacobson, I.Kirsch, R.N. Schyner, B.H. Nam, L.T. Nguyen, M. Park, A.L. Rivers, C. McManus, E. Kokkotou, D.A. Drossman, P. Goldman and A.J. Lembo (2008). Components of placebo effect: randomised controlled trial in patients with irritable bowel syndrome. *BMJ* April 3. (http://www.ncbi.nlm.nih.gov/sites/entrez) Viewed April 9, 2008

Kirkpatrick, I. and M.M.Lucio (Eds.) (1995). *The politics of quality in the public sector.* London and New York: Routledge.

Launsø, L. (1989). "Integrated medicine in practice—a challenge to the health-care system." *Acta Sociologica* 32,3: 237-251.

Launsø, L. (1996). *Det alternative behandlingsområde.—Brug og udvikling; rationalitet og paradigmer* [The alternative treatment field. Use and development; rationality and paradigms]. Copenhagen: Akademisk Forlag.

Launsø, L. and D.E. Gannik.(2000). "The need for revision of medical research designs." In D.E. Gannik and L. Launsø (eds.). *Disease, Knowledge and Society.* Copenhagen: Samfundslitteratur.

Launsø, L. and N. Haahr (2007). "Bridge building and integrative treatment of persons with Multiple Sclerosis. Research-based evaluation of a team building process." *Journal of Complementary and Integrative Medicine (JCIM)* (Accepted for publication).

Launsø L.; I. Henningsen, J. Rieper, H. Brender, F. Sandø and A. Hvenegaard (2007). "Expectations and effectiveness of medical treatment and classical homeopathic treatment for patients with hypersensitivity illnesses—one year prospective study." *Homeopathy* 96:233-242.

Li, J., C. Johansen, H. Brønnum-Hansen, E. Stenager, N. Koch-Henriksen and J. Olsen (2004). "The risk of multiple sclerosis in bereaved parents. A nationwide cohort study in Denmark." *Neurology* 62(March): 726-729.

Mann, D.; S. Gaylord and S. Norton (2004). Moving toward integrative care: Rationales, models and steps for conventional-care providers. *Complement Health Practice Review* 9:155-172.

Marrie, R.A., O. Hadjimichael and T. Vollmer (2003). "Predictors of alternative medicine use by multiple sclerosis patients." *Multiple Sclerosis* 9,5: 461-466.

Mayne, J. and O. Rieper (2003). "Collaborating for public service quality: The implications for evaluation." Chapter 5 in A. Gray, B. Jenkins, F. Leeuw and J. Mayne (eds.), *Collaboration in Public Services: The Challenge for Evaluation,* New Brunswick (USA) and London (UK): Transaction Publishers.

Melucci A. (1989). *Nomads of the present—social movements and individual needs in contemporary societies.* London: Hutchinson Radius.

Mulkins, A.L. and M.J. Verhoef (2004). "Supporting the transformative process: experiences of cancer patients receiving integrative care." *Integrative cancer therapies* 3,3: 230-237.

Mulkins, A.L., J. Eng and M.J. Verhoef (2005). "Working towards a model of integrative health care: Critical elements for an effective team." *Complementary Therapies in Medicine* 13:115-122.

Mulkins, A.L., M. Verhoef, J. Eng, B. Findlay and D. Ramsum (2003). "Evaluation of the Tzu Chi Institute for Complementary and Alternative Medicine's Integrative Treatment Program." *Journal of Alternative Complementary Medicine* 9,4: 585-592.

Munger, K.L., S.M. Zhang, E. O'Reilly, M.A. Hernan, M.J. Olek, W.C. Willett and A. Ascherio (2004). "Vitamin D intake and incidence of multiple sclerosis." *Neurology* 62: 60-65.

Naschold, F. (1995). *The modernisation of the public sector in Europe. A comparison perspective on the Scandinavian experience.* Helsinki: Ministry of Labour.

Ostenfeld-Rosenthal, A. (2007). "Symbolsk healing 'embodied'. Krop, mening og spiritualitet i danske helbredelsesritualer." [Symbolic healing 'embodied'. Body, meaning and spirituality in Danish healing rituals] *Tidsskrift for Forskning i Sygdom og Samfund* 6:129-150.

Paterson, P. (2004). "Seeking the patient's perspective: A qualitative assessment of Euro-Qol, COOP-WONCA charts and MYMOP." *Quality of Life Research* 13: 871-881.

Paterson, P. and N. Britten (2000). "In pursuit of patient-centred outcomes: A qualitative evaluation of MYMOP, Measure Yourself Medical Outcome Profile." *J Health Ser Res Policy* 5: 27-36

Paterson, P. and N. Britten (2004). "Acupuncture as a complex intervention: A holistic model. *The Journal of Alternative and Complementary Medicine* 10,5: 791-801.

Pawson, R. (2006). *Evidence-based policy. A realist perspective.* London: SAGE Publications Ltd.

Pollit, C. and G. Bouckaert (Eds.) (1995). *Quality improvement in European public services/concepts, cases and commentary.* London: Sage.

Riise, T., M.W. Nortvedt and A. Ascherio (2003). "Smoking is a risk factor for multiple sclerosis." *Neurology* 61: 1122-4.

Sandaunet, A.-G. (2008). *Keeping up with the new health care user: the case of online-help groups for women with breast cancer.* Tromsø: University of Tromsø, Institute of Sociology, June.

Scherwitz, L., W. Stewart, P. McHenry, C. Wood, L. Robertson and M. Cantwell (2003). "An integrative medicine clinic in a community hospital." *American Journal of Public Health* 93(4):549-552.

Scocozza, L. (2000). "The randomized trial. A critique from the philosophy of science." In: D.E. Gannik, and L. Launsø (eds.), *Disease, knowledge and society*, Copenhagen: Forlaget Samfundslitteratur.

Storr, L.K. (2002). *Brugeren og brugen af alternativ behandling. En populationsbaseret undersøgelse af scleroseramtes brug af alternativ behandling i Danmark* [The user and use of alternative treatment. A population-based investigation of alternative treatment use among Multiple Sclerosis patients in Denmark]. PhD Thesis (unpublished).

Sørensen, P.S., M. Ravnborg and A. Jønsson (eds.) (2004). *Dissemineret sklerose. En bog for patienter, pårørende og behandlere* [Multiple Sclerosis. A book for patients, relatives and practitioners]. Copenhagen: Munksgaard.

The Danish Multiple Sclerosis Society (2002). *Livet med Sclerose i Danmark* 2002 [Living with Multiple Sclerosis in Denmark 2002]. Unpublished.

Touraine, A. (1981). *The voice and the eye.* New York: Cambridge University Press.

Wampold, B.E. (2001). *The Great psychotherapy debate: Models, methods, and findings.* LEA's Counseling and Psychotherapy Series.

Part III

Using Evidence

7

Results Management: Can Results Evidence Gain a Foothold in the Public Sector?

John Mayne

Introduction

Results management has become common in the public and not-for-profit sectors (Mayne and Zapico, 1997; Moynihan, 2006; Norman, 2002; OECD, 2005; Pollitt and Bouckaert, 2000). I will be using the term *results management* to cover a variety of terms used in the literature, such as result-based management, managing for results, managing for outcomes, outcome-focused management, performance-based management and performance management. Reflecting on several decades of public sector reform in member countries, the OECD (2005: 56) concludes: "The performance orientation in public management is here to stay. The benefits of being clearer inside and outside government about purposes and results are undeniable."

The intent behind this movement is to deliberately measure what results—i.e., outputs and outcomes—are brought about by policies, programs and services, and to use that information to help better manage public funds and to better report on the use of those funds. It is an approach that tries to bring information on results and on expected results to bear on managing.

Despite widespread efforts at implementing results management, the challenges are many (Binnendijk, 2001; Curristine, 2005; Government Accounting Office, 2002; Kamensky and Morales, 2005; Mayne, 2007b; Moynihan, 2006; Perrin, 2002, 2006). And in particular, getting managers and others to actually use the results information produced is not straightforward, as many have noted (Johnson and Talbot, 2007;

Moynihan, 2005; Pollitt, 2006; ter Bogt, 2004; Weiss, 1998). In this chapter, I want to explore more carefully, why this may be so.

A parallel movement that this book discusses is the approach of evidence-based policy (Chalmers, 2005; Davies, Nutley and Smith, 1999; Learmont and Harding, 2006; Nutley, Walter and Davies, 2003, 2007; Pawson, 2002a, 2002b; Sanderson, 2002, 2004). A variety of terms are used here also, such as evidence-based management, evidence-based practice, and evidence-based policy and practice. I shall mostly use the generic term *evidence movement* to cover these approaches. For some, the evidence movement implies bringing the strongest possible scientific evidence to the table, such as from randomized controlled trials (The Council for Excellence in Government, 2008). I take a broader perspective and look at evidence-based approaches as systematic efforts to bring empirical evidence to the management table, and thus that results management is indeed an evidence-based management approach. The underlying research question is the same: To what extent can empirical evidence be made a part of managing in the public and not-for-profit sectors?

There are likely many reasons for the less than hoped for use of results information and no single cause, but perhaps results management has not been thought through well enough in terms of how in the real world of managers it is supposed to work. Just how and when is which kind of evidence supposed to be used? What needs to be in place in an organization to ensure results information is valued and seen as essential for good management? These are the questions examined in this chapter. Part of the aim is to bring the evidence movement thinking to bear on results management.

First, a short overview of the evidence-based movement will be provided and key terms defined. Then, traditional management will be compared with results management to highlight the differences and in particular, what is expected of managers that are managing for results. The theory behind how results information is supposed to be used by management will be examined, with a view to identifying what conditions are likely to enhance its use in decision making. In the final sections of this chapter, requirements and approaches to building and maintaining a results culture in organizations will be discussed and directions for strengthening results management, building on evidence-based policy insights will be suggested.

The Evidence Movement

Bringing empirical information to bear on decision making in the public sector is not new. In the past 30 years, evaluation and performance measurement have become widely-used tools in public administration. Williams (2003), indeed, discusses performance measurement in the early twentieth century. These are approaches to measuring aspects of the performance of an organization, or its policies, programs and services, in order to provide empirical evidence on how well things are working. Are objectives being met? Are outputs being delivered as planned? Are there unintended impacts resulting from the activities being undertaken? Data is gathered and analyzed, and the resulting information provided to decision makers. All such efforts fall into, I would argue, a broad category of bringing empirical evidence to the decision making table.

Here I am using the term "evidence" in its everyday meaning: "the available facts, circumstances, etc. supporting or otherwise a belief, proposition, etc. or indicating whether or not a thing is true or valid" (*Canadian Oxford Dictionary*). More simply, perhaps, *evidence* in the context of this book is information pertinent to a conclusion being drawn, such as a decision. But interest here is not in all kinds of such information, but in *empirical* information, i.e., information "based on observation and experiment, not on theory" (*Concise Oxford Dictionary*). In this chapter, I am including under the "empirical" label findings derived from "systematic reviews" of a set of other empirical studies of the kind produced by the Cochrane Centre (http://www.cochrane.org), the Campbell Collaboration (http://www.campbellcollaboration.org) and the Evidence for Policy and Practice Information and Coordinating Centre (http://eppi.ioe.ac.uk/cms).

It is well acknowledged that there are significant other types of information that are used in public administration: personal experience, hearsay, beliefs, etc. Empirical information competes for attention in public administration with these other forms of information.

The broad phenomenon being considered, then, is using empirical evidence in public administration. This chapter is looking at results management as a widely adopted approach to bring empirical evidence to bear on public and not-for-profit sector decision making.

This discussion on evidence has so far made no mention of quality, of what counts as evidence. Whether evidence is good or bad, useful or not, are separate questions from what evidence is. The introductory chapter in this book made references to Thomas' (2004) three criteria for judging evidence: relevance, sufficiency and veracity. And the use of evidence depends on numerous factors, many of which are based on issues outside the evidence itself, such as the presentation of the evidence, its availability and the beliefs held by those who might use the evidence.

The evidence movement frequently uses "evidence" in a much more restrictive way than it is used in everyday discussion. It is often seen as pushing for one type of evidence to be used—that produced through randomized controlled trials, seen as the "gold standard"—and not others. For example, see The Council for Excellence in Government (2008). Nutley, Walter and Davies (2007) call this the narrow definition of evidence-based policy and practice. The expectation is that with such strong evidence, decisions can be safely taken, essentially determined by the evidence.

On the other hand, many authors, such as Learmonth and Harding (2006), argue for a more heterogeneous supply of evidence. Similarly, Nutley, Walter and Davies (2007) take a much broader view of what counts as evidence. In their view, "fit for purpose" acts as the main criterion for what counts as good evidence" (p. 13). The introductory chapter in this volume argues that "there is a variety of types of evidence that might be more or less fit for different purposes." I concur. There is a range of types of empirical evidence and a range of quality of that evidence. As I have argued elsewhere (Mayne, 2001: 6):

> Measurement in the public sector is less about precision and more about increasing understanding and knowledge. It is about increasing what we know about what works in an area ... [W]e can almost always gather additional data and information that will increase our understanding about a program and its impacts, even if we cannot "prove" things in an absolute sense.

Of course, quality matters and evidence brought to the table should clearly state its limitations. And the measurers professionally should be aiming to gather as credible evidence as possible in a particular situation. Clearly the context does matter. In, for example, drug approvals and in many other health areas, one would expect randomized controlled trials-type evidence to dominate, not the views of officials based on hearsay. The degree of credibility of the evidence that is needed depends on the contestedness of the area/issue in question. We need feasible and timely evidence, "fit for purpose."

Results management aims to bring to the management table available results information on performance issues, but that evidence is rarely definitive: it is often contested. It is used to inform not determine decision making.

A final note. *Results information*, as I am using the term, includes not only empirical evidence on results that have been observed, but also the assumptions and understandings behind programs and interventions. As Nutley, Walter and Davies (2007: 299) argue in respect of social research more generally, "Theorizing, reconceptualizing and inference drawing can all add insights, and can contribute to problematizing of current ways of thinking, and gradual reframing over time." As I will argue shortly, the theory of change and its assumptions behind a program or intervention is an important aspect of results management.

What is Really Different with Results Management?

Organizations are asking their managers to manage with a focus on results. Results management is seen as essential to improving performance, but managers might reasonably ask:

> All this results focus is fine, but what is really *different* for me if I undertake results management? What do I have to do differently from what I do now?

This is a legitimate question and needs to be addressed. Here, three models of program management will be presented and compared:

- traditional management,
- managing for outputs, and
- managing for outcomes.

Managing for outputs and managing for outcomes are forms of results management, with managing for outputs often seen as a precursor, but an approach that has been pursued in several countries as an end in itself—the United Kingdom with Next Steps Agencies (Goldsworthy, 1991) and in New Zealand with its early public sector reform models (Boston, Martin, Pallot, and Walsh, 1991). In practice, there can be a gradation between output and outcome-focused management.

Without entering into the theories of management, five elements of management will be identified to frame the discussion:

- planning
- implementation/delivery
- monitoring/measuring
- adjusting/correcting/learning
- accounting for/reporting on performance

The intent for each of the three management models is to set out what each entails for managers and to do so in a realistic light, not setting up straw men! The focus is on the role results thinking and information plays or doesn't play in managing, not on all aspects of managing, i.e., not on, for example, financial and human resource management. In each case, the elements of good results management are set out. Diamond (2005) discusses the differences among traditional budgeting, output-based budgeting and outcome-focused budgeting.

Traditional Program Management

Given the attention on results in management over many years now, "traditional" program management only weakly focused on results may be hard to find in a pure form, but elements of it may well be recognized. Table 1 sets out the key elements of good (traditional) program management.

Managing for Outputs

Managing for outputs introduces two key concepts in results management:

1.   the use of performance expectations in managing and reporting, and
2.   the deliberate use of the performance information to learn and adjust delivery.

Table 2 outlines the elements of good managing for outputs.

As can be seen, managing for outputs builds on the elements of traditional management, but adds several new aspects to a manager's job:

Table 1
Elements in Traditional Program Management

Planning
- establishing what impacts are sought
- determining the activities to be carried out and the resources required

Delivery
- attention to ongoing delivery—managing resources and activities

Monitoring
- measuring resources used, workloads, activities undertaken and outputs delivered

Adjusting
- modifying delivery as required when problems arise

Accounting for performance
- reporting on resources used in relation to budget (financial reporting)
- reporting on level of effort, outputs delivered and their costs (activity reporting)

Table 2
Elements in Managing for Outputs

Planning (output-based)
- establishing what impacts are sought
- determining the activities to be carried out and the resources required
- setting meaningful (realistic but challenging) output and cost expectations/targets
- developing a strategy for measuring key outputs through ongoing indicators

Delivery
- attention to ongoing delivery—managing resources and activities

Monitoring and Analysis
- measuring resources used, workloads, activities undertaken and the outputs delivered
- analyzing the data and information gathered in light of the expectations set

Adjusting and Learning
- modifying delivery as required when problems arise
- on an ongoing basis, using the monitoring information to improve delivery
- on a periodic basis, reviewing the output and cost expectations and the measurement strategy as to continued relevance and cost, and how the monitoring information is being used

Accounting for performance
- reporting on resources used in relation to budget (*financial reporting*)
- reporting on level of effort and costs (*activity reporting*)
- reporting the attainment of performance expectations/targets (outputs) and their costs (*performance reporting*)

The underlined elements highlight the differences from Table 1 on traditional program management.

- setting out specific expectations on the outputs to be delivered at what cost,
- identifying how progress towards the expectations is to be measured,
- analyzing the monitored data in light of expectations,
- using the output information to improve delivery,
- reviewing what evidence is being gathered, and
- reporting against the output expectations.

Managing for outputs is fairly widespread and is often what managers fall back on in the face of difficulties in managing for outcomes. It requires a modest effort to gather evidence on the direct results of the program activities—the outputs delivered—typically through monitoring of a number of performance indicators which track key outputs.

Managing for Outcomes

Managing for outcomes goes beyond just focusing on outputs and asks in addition that managers:

1.   understand the theory of change embodied in the program and the risks to it,
2.   measure the various outcomes the program is contributing to, and
3.   seek to understand why the outcomes are resulting from or influenced by the outputs being delivered.

Table 3 sets out what good managing for outcomes entails and shows the differences with traditional program management.

When outcomes become part of the management regime, in addition to measuring those outcomes, managers are asked to understand the theory behind their programs and its strengths and weaknesses—i.e., the risks faced by the program in achieving its intended results—and to undertake measurement and research activities to better understand why the outcomes observed are being produced. The OECD-DAC *Managing for Development Results Sourcebook* (2006: 9) puts it well:

> Results-based management asks managers to regularly think through the extent to which their implementation activities and outputs have a reasonable probability of attaining the outcomes desired, and to make continuous adjustments as needed to ensure that outcomes are achieved.

This includes examining to what extent the program's activities have contributed to those outcomes, as opposed to other influencing factors—the cause-effect question—and looking for unintended outcomes. Managing for outcomes requires actively seeking reliable evidence on outcomes, as well as outputs. It should encourage an active interest in evaluation and research on the impacts of programs and a desire to learn from that evidence.

Programs vary considerably in their complexity and in how well their theories of change are understood and accepted. Managers of programs that are delivering a straightforward service to beneficiaries—such as information on income tax regulations, checks to eligible people, collection of garbage, delivery of electricity or water—do not need to spend much time focusing on outcomes, other than perhaps client satisfaction issues associated with the service being delivered. Managing for outputs is probably quite adequate in these cases. There are outcomes resulting

Table 3
Elements in Managing for Outcomes

Planning (results-based)
- establishing what outcomes are sought
- knowing and questioning the theory of change behind the program, the evidence for it, and the risks to it
- determining the activities to be carried out and the resources required
- setting meaningful expectations for the results chain (outputs, outcomes/impacts)
- developing a strategy for measuring key outputs and outcomes through ongoing indicators, periodic evaluations and other studies

Delivery
- attention to ongoing delivery—managing resources and activities

Monitoring, Analysis and Evaluation
- measuring resources used, workloads, activities undertaken, and outputs delivered

Results monitoring and evaluation:
- measuring and analyzing actual outcomes/impacts in light of expectations
- tracking other influencing factors
- assessing the contribution the program is making
- undertaking evaluations and other research as needed to further understand why the outcomes are occurring and if there are unintended outcomes
- analyzing this performance information in light of the expectations set

Adjusting and learning
Delivery improvement
- modifying delivery as required when problems arise
- on an ongoing basis, using the performance information for improving delivery

Design improvement
- on a periodic basis, using the performance information to improve program design, relevance and effectiveness
- modifying the accepted theory of change and the assessment of risk as required

Measurement strategy improvement
- On a periodic basis, reviewing the performance expectations and the measurement strategy as to continued relevance and cost, and how the performance information is being used

Accounting for performance
- reporting on resources used against budget (financial reporting)
- reporting on level of effort and costs (activity reporting)
- reporting on extent of attainment of performance expectations (outputs, outcomes/impacts) and costs (*performance reporting*)
- reporting on contribution made by the program to the observed outcomes (*performance reporting*)

The underlined elements indicate the differences from Table 1 on traditional management.

from the delivery of the services, but in many cases these are not the objectives of the program per se. Interest in outcomes, such as monitoring if there are unintended outcomes occurring associated with the service being provided are usually seen as the responsibility of others in the organization, perhaps a policy research or evaluation group.

In many other cases, the manager is indeed trying to bring about important outcomes through the delivery of specific goods and services. And often, the theory of change of the program is not well understood and often contested; evidence is needed to better understand how to best bring about or contribute to the desired outcomes. In these cases, managers are managing for outcomes, and the elements in Table 3 are pertinent to their work.

The differences from the traditional program management model are several and summarized in Table 4. In essence, in managing for outcomes managers are expected to actively seek evidence on the extent to which the theory of change of the program is being verified and to use that information to improve the results sought.

Having noted the differences, it is also important to know *what doesn't change with managing for outcomes*:

- on a day-to-day basis, managers still will be paying attention to delivery, managing what they can control (resources, activities, and outputs) and correcting problems as they arise.

So while the bulk of the time spent by a manager managing for outcomes will still be on day-to-day operational issues, the manager will have specific performance expectations in mind and know how and why the program activities are expected to lead to the achievement of the impacts. The manager will be deliberately seeking out evidence on how well the program is doing against those expectations and on the theory of change behind the program. The manager will be adjusting delivery in light of this evidence, and from time to time, based on the available evidence, will review and reconsider the program, its implementation and its design as to what needs changing, revising the theory of change are required.

I would argue that, while each of the elements listed in Table 4 are challenging and in need of improvement in most organizations, the main overall weakness in results management is in the use of results information for decision making and learning. A look at the planning and reporting documents of many organizations in the public sector and not-for-profit sectors show clearly a results orientation, with performance

Table 4
Specific Expectations for Managers Managing for Outcomes

In planning:
Understand the theory of change. Knowing and questioning the theory of change, the evidence for it and the risks to it—why the program is believed to contribute to the outcomes sought.
Set out performance expectations. Setting meaningful expectations/targets for key aspects of the theory of change (outputs and outcomes/impacts).

In implementation:
Measure and analyze results, and assess contribution. Gathering evidence and information on key outputs, outcomes/impacts, unintended outcomes and other influences occurring, assessing that information in light of the expectations set, and assessing the contribution being made by the program to the observed outcomes/impact.

In decision making and learning:
Deliberately learn from evidence and analysis. Using this evidence and analysis to adjust delivery and, periodically, modify or confirm program design and delivery. Reviewing periodically the evidence being gathered as to its continued relevance, usefulness and cost.

In accounting for performance:
Reporting on performance achieved against expectations. Reporting on the accomplishment of output and outcome expectations, and on the contribution being made by the program—what difference it is making.

expectations set out, sometimes with targets, and performance reporting on the results achieved, implying some level of monitoring of results being undertaken. The planning, monitoring and accounting of performance may be quite weak or in need of improvement—as many national audit offices have pointed out (Auditor General of Canada, 2000a, 2000b; Australian National Audit Office, 2007)—but the effort is being made.

It is in the area of the use of results information for decision making and learning that many observers say is the major weakness of results management (Moynihan, 2005; Pollitt, 2006). The second biggest challenge is working with theories of change. Many managers today have had some exposure to results management seen as developing a few indicators that are then tracked. Results chains and theories of change are just beginning to be seen as essential to outcome management.

The question then, is what can be done to foster better use of results information including theories of change in decision making and learning? This, of course, is the same basic question that the evidence movement is asking.

A Theory of Change for Using Information on Results

I want to return now to the question: just how is information on results expected to be used in decision making and learning? We saw that this expectation of managers—and likely similarly for senior managers—is the weakest element of results management in many organizations. But it is clearly a rather large gap since a key intent is to make use of information that is deliberately gathered on results, as a way to improve the management of public authorities and funds.

I want to look at just why or exactly how this "use" of results information is expected to occur. What conditions are needed for information—evidence—on results to be used? What indeed is the theory of change behind this key expectation of results management?

Realistic Use of Evidence in an Organization

First, we need to agree on what is meant by "use." Much has been written about using results information—evidence—in decision making. For example, in the evidence movement literature, see Nutley, Walter and Davies (2003 and 2007), Learnmont and Harding (2006), Pawson (2002a, 2002b), Sanderson (2002), and Walter, Nutley and Davies (2005). There is a perhaps an even larger evaluation and results management literature on the utilization issue (for example, Cousins, Goh, Clark and Lee, 2004; Feinstein, 2002; Leeuw, Rist and Sonnichsen, 1994; Leviton, 2003; Mark and Henry, 2004; Moynihan, 2005; Rist and Stame, 2006; Sridharan, 2003; Weiss, 1998). Others have discussed what "use" means (Caracelli and Preskill, 2000; Marra, 2004; Valovirta, 2002; Vedung, 1997). Nutley, Walter and Davies (2007) devote Chapter 2 of their book to discussing what it means to use research evidence.

Use is often in practice taken to mean that we should expect decisions to reflect the results information; that robust findings of evaluations, for example, should be implemented. I would argue, rather, that we should be more modest in our expectations. Results information is but one type of information that is usually considered in decision making (Chalmers, 2005; Nutley, Walter and Davies, 2007; Sanderson, 2004). Results information plays a role as an aid to decision making. In this light, results information is "used" when it is seriously considered in discussions and debates surrounding decisions, i.e. the information gets debated and discussed. Results information should inform decisions, not dictate them.

Formal research and evaluation evidence may well provide guidance on "what is likely to work" but this will need to be assimilated with practical wisdom in coming to decisions on policies or actions that are appropriate in particular circumstances, taking into account relevant ethical-moral considerations (Sanderson, 2004: 371).

Although phrases such as "evidence-based policy" continue to be widely used, a number of authors have noted that a better term might be "evidence-informed policy." As far back as 1981, Leviton and Hughes define utilization in a similar manner. Joyce (2005) discusses what he prefers to call "performance–informed budgeting" in the United States. He notes that budgets are "appropriately influenced by other (nonperformance) concerns such as relative priority, unmet needs, and equity concerns" (p. 98). Nutley, Walter and Davies (2003: 126) suggest that evidence-informed or evidence-aware policy might be better terms to use, as does Chalmers (2005). Sanderson (2002: 4) discusses a variety of ways that evidence can inform policy making.

Another important aspect of the context here is that I am concerned—as is results management—with use in an organization on an ongoing or routine basis, not use in a special one-off case. On an occasional basis, results information such as from an evaluation may get used quite directly as the result of the alignment of various factors. This is quite different from routine use, which results management aims for. In the organizational setting of seeking routine use, there are opportunities for learning how to enhance such use, but also considerable challenges in ensuring that evidence on results is seen as part of good managing. The evidence movement literature is not always clear as to which type of use is being discussed, although much of it does seem to be discussing the one-off use of particular research evidence.

I have framed the discussion so far as talking about what the utilization literature calls *instrument use* of results information where specific decisions are influenced by results information (Leviton and Hughes, 1981; Sanderson, 2002). Some argue such use is unrealistic and that we should rather talk about *enlightenment use* (Weiss, 1998), whereby, over time and being exposed to a variety of results information, individuals (and thereby organizations) adjust their views about a problem they are aiming to improve. They learn from using results information. Here more clearly, the aim is to have results information inform, this time to inform thinking and learning about issues. I therefore see the issue of using results information in either an instrumental or an enlightenment mode as quite similar, namely to inform those involved.

The differences have mainly to do with timing, although even here the differences may not be as pronounced as often thought, since many significant decisions in a bureaucracy actually occur over a period of time. Many decisions are not made at a single meeting, rather at different times evidence is presented and debated, and reflection occurs— enlightenment or conceptual use. Further debate and perhaps additional evidence is brought to the table, and eventually a decision is taken—instrumental use.

A Theory of Change for Using Results Information and Evidence

With this as the context, Table 5 sets out basic elements of a suggested theory of change for the use of results information in results management.

Each of the assumptions in the theory of change outlined in Table 5 hides numerous sub-assumptions and requires additional explanation and discussion, particularly Assumption 6. In one sense, several of the assumptions—relevance, availability (timeliness), understandability (clarity, transparency), reliable and credible—are not surprising; they represent what is seen as characteristics of good performance information. See for example Schwartz and Mayne (2005).

The model outlined in Table 5 has the appearance of what Nutley, Water and Davies (2007: 50) call "stage" models of use, models that are basically linear and, they argue, do not reflect the more iterative nature of much use of information. However, by incorporating both instrumental and conceptual use, the model without being explicit, actually allows for a variety of iterative learning to occur. Often, as was noted, conceptual

Table 5
A Theory of Change for Using Results Information

Assumption 1. If there is an issue or decision to be addressed, AND
Assumption 2. If relevant results information is available, AND
Assumption 3. If the information is made available at times when it can be considered, AND
Assumption 4. If the information is understandable, AND
Assumption 5. If the information is seen as reliable and credible, AND
Assumption 6. If there is interest in results information by those involved, THEN,

the information will be seriously considered in making the decision (instrumental use) or addressing the issues (conceptual/enlightenment use)—it will inform the issues and/or decisions, and learning takes place.

use occurs around an issue before actual decisions are taken, which themselves may be informed by newer results evidence.

Each assumption in the model will be discussed, before addressing the questions: Is the theory complete? Are the assumptions sufficient? Are they necessary? Does it matter?

Assumption 1: There is an Issue or Decision to be Addressed

The point here is that there has to be a context for the results information to be used, either a decision to be made or issues for which people are seeking clarification and greater understanding. Just providing information that looks interesting to people is unlikely to result in much use.

What then are the various situations where results information might inform management? I suggest:

- Decision making, namely in
  - Policy formulation
  - Planning decisions (setting goals and expected results; designing implementation approaches)
  - Programming and reprogramming decisions (adjusting outputs; adjusting short term goals)
  - Policy and program termination
  - Operational decisions (keeping activities/outputs on track; adjusting inputs)
  - Fire-fighting
- Learning taking place, where decisions per se are not being taken:
  - Learning takes place cumulatively over time, or
  - At learning events (Barrados and Mayne, 2003; Mayne and Rist, 2006; Moynihan, 2005)

These are all, of course, different situations requiring different types of information, suggesting the need for some clarity as to the context for the expected use of results information. Lindquist (2001), in discussing how evaluations might influence policy, identifies four types of decision situations: routine, incremental, fundamental, and emergent, all requiring different types of information. Furubo (2006) discusses evaluations' influence on fundamental policy decisions, maintenance decisions and operative decisions. It is important to know and understand the decision context into which empirical evidence is being introduced.

Assumption 2: Relevant Results Information is Available

This can be a challenging assumption, especially in a decision making context. Decision-makers may not be quite sure just what information they do want; circumstances can change quickly so that new issues emerge overshadowing what was thought to be relevant. What circumstances enhance the chances that this assumption holds, i.e., that the results information is relevant?

Key issues have been identified a priori. For decisions that can be planned in advance, the kinds of results information that could be useful should be able to be identified. Policy decisions, for example, often have a long gestation period. Fire-fighting decisions, on the other hand, leave little time for planning. If the information does not happen to be at hand, decisions are made anyhow.

To some extent when managers set out performance expectations and targets for their programs, or indeed, a theory of change for their programs, they are identifying key issues. That is, results-based planning documents do set out some of the key issues of interest, but probably not all. If the key issues to be addressed have been identified, and there is time, then relevant results information could be made available.

The right type of information is available. But even if there is a reasonable idea about the issues in need of further information, getting the right relevant information is often not easy. There are different types of results information—more accurately results knowledge—possible:

- *Basic results data*—data on the occurrence or not of outputs and outcomes.
- *Results analysis*—analysis of the results data, such as comparisons with expected results, explanation of differences with expected results, trends over time, etc.
- *Assumptions and rationale analysis*—analysis of the assumptions and understanding of the theory of change behind the program in question, including the ongoing rationale for the program.
- *Causal claims*—claims and statements about the extent to which the program activities have indeed made a contribution to the observed results; on whether the program is making a difference
- *Action-oriented analysis*—statements about the implications of the results data and information for decision making.

Others have presented different frameworks of types of knowledge. See, for example, Leeuw (2006), Lindquist (2001), and Nutley, Walter and Davies (2003).

Much performance measurement information that is made available is simply basic data on the volume of outputs and outcomes observed, with perhaps some variance analysis. Often this is not very rich data, and perhaps of limited usefulness. Many contexts need some information on causal claims and suggestions for future action. Perrin (2006: 40) makes the point that what decision-makers usually want is information to help them make decisions about the future. In such contexts, basic results data per se may not be "relevant." As the OECD (2005: 75) notes:

> Performance indicators and targets provide a snapshot of performance in time. They do not provide a guide to future performance nor do they explain why a target has been achieved.... Unlike targets, evaluations can explain the results of a policy or program and what changes will improve its performance.

Sanderson (2004: 372) is even more critical of the use of indicators and targets:

> "... the "target culture" is not conducive to an approach to professional practice that encourages the search for evidence and its application in conjunction with practice wisdom...."

To get relevant information, either managers need to specify in advance what they are likely to need, or those producing the information need to know well the policy cycle in the organization (Mayne and Rist, 2006; Sanderson, 2004)—the context within which programs are working—and perhaps through discussion, identify the likely information needs of the organization. In practice, hopefully, both these conditions can be worked on.

Relevance to public issues. If the information is relevant to current public debate, then it is more likely to be considered by decision makers. Leeuw, van der Knaap and Bogaerts (2007) describe a case where synthesis information on violent crime in the Netherlands was use immediately by policymakers. A key reason, they ague is because the topic of violent crime was high on the public agenda.

Assumption 3: The Information is Made Available in a Timely Fashion

This is a reasonably clear assumption, although hard to deliver on for instrumental use (Pawson, 2002a). The information has to be

- available prior to a decision being taken, and
- with enough time available to understand and consider the information (Mayne and Rist, 2006).

For all but the fire fighting type decisions, the timing of the decision is often known in advance. Operational reviews (annual, monthly or weekly), rethinking programming options, planning decisions and policy formulation are examples. In such cases, those producing results information should be able to deliver on time.

Timing may be less of an issue for enlightenment use. Ongoing learning contexts are usually more flexible. If an issue is being discussed and debated over a period of time, results information needs to be available during that period, with ample time for reflection. Learning events can be planned in advance with specific information needs that should be able to be met.

Results information would be available for fire-fighting decisions only if it were part of an ongoing results information system or data base that could be queried at any point in time.

Utilization in the Leeuw, van der Knaap and Bogaerts (2007) case study was helped by the ability of the study team to frequently interact with the decision makers to explain their findings, including preliminary findings. They were able to bring the relevant findings to the attention of decision-makers.

Assumption 4: The Information is Understandable

Again, a fairly clear assumption, but often not met. The results information has to be communicated in a form that is digestible, clear and understandable, and this will differ depending on the audience. Operational managers might be quite comfortable with tables of results data, while senior managers and parliamentarians probably want short summaries, or even verbal reports set in context.

The challenge is to take the time to know how best to communicate the results information to the specific audience. See for example, Torres, Preskill and Pointek (2005).

Assumption 5: The Information is Seen as Reliable and Credible

Producing results information that is seen as reliable and credible is not straightforward. What counts as evidence is a key question in the evidence movement. In health areas, for example, randomized controlled trials are seen by many as what is required for good evidence (Davies, Nutley and Smith, 1999). Much has been written about the quality of results data and information (Schwartz and Mayne, 2005). Obviously, following good social science methods helps. But in an organization, the credibility of those producing the results information also matters. Establishing a record over time of delivering reliable and credible information would go a long way to ensuring this assumption holds.

In their case study, Leeuw, van der Knaap and Bogaerts (2007) not only point to the credibility of the evaluation findings as a contributing factor in getting the study used, but also note that it was produced by a quasi-independent unit within the Ministry of Justice.

Assumption 6: There is Interest in Results Information by Those Involved

This may be the most demanding assumption—and the main one this paper explores—as it gets into motivation and the mindsets of those involved. There are many reasons why, even in the face of the "right" results information—relevant, timely, understandable, reliable and credible—some may not want to seriously consider the information, such as:

- those involved don't "trust" empirical information, believing rather in their own experience and knowledge,
- the information doesn't fit their ideological base,
- the information is inconvenient for power struggles, budgets, careers, and/or
- the implications arising from the information challenge current practices and require more willpower to change than there is.

In such contexts, assumption 6 won't hold and results information is unlikely to be seriously considered.

More interestingly, in what contexts would the assumption likely hold? Interest in results information is likely when, for example:

The evidence is seen as convenient. This might be the case when the evidence supports ideological positions; or those providing and using the information see it in their interest to do so. "Kogan (1999) argues that governments will seek to legitimize their policies with references to the notion of evidence-based decision making but use research evidence only when it supports politically-driven priorities" (Sanderson, 2002: 5). This is not the context for use one would like to see, but it exists. What might be said for it is that having used evidence in such a case, it may be more difficult to not use similar information in the future when the information challenges beliefs. But it is also not a context that would sustain the use of results information, since the next results information may not be seen as convenient.

The need for evidence is seen as essential or at least worthwhile. When a culture of seeking evidence is not widespread, the need for evidence in specific cases might occur when, for example

- senior officials call for results evidence and will proceed to make decisions even if the results information is not available;
- special requests are made for specific results information for important discussions rather than just using routine data; or
- there is an influential supporter of the results information, perhaps a respected colleague who has shown that such information can be useful—a results champion—and hence others may be willing to "give it a try."

Again, this context will more likely lead to episodic interest in results information than sustain the use of results information in an organization over time. Nevertheless, in most organizations, sustained use would likely only occur after some time and experience with results management is acquired.

A culture of seeking evidence exists. Due to leadership or experience over time, expecting results information including empirical evidence to support managing and decision making can become the norm in an organization or sub-unit in an organization. In such contexts, the lack of results information would be seen as incompetence. In this context, routine use of results information is likely.

I conclude that for everyday interest in results information in an organization, what is needed is an organizational culture where results information is valued and seen as a normal part of good management. This is not a surprising finding. Many have noted the need for a results-oriented culture (Curristine, 2005; Kamensky and Morales, 2005; Mayne, 2007c; Moynihan, 2005; Nutley, Walter and Davies, 2007; Perrin, 2006; World Bank-OED, 2005).

This though begs the further question of what is needed to build, support and sustain such a culture?

Building an Organizational Culture for Results Management

The suggested theory of change for using results information requires an interest in results information for the information to be used. And, on an ongoing basis, such an interest needs a climate in the organization where evidence is valued and seen as essential to good management. The question left unanswered was how such a culture and climate could be built and fostered.

Earlier, Table 4 outlined what results management in practice requires of a manager, over and above "regular" day-to-day management. Again, the unanswered question was how to get managers to undertake these tasks? What kinds of incentives and supports could foster a results culture and move an organization towards results management?

What is a Culture of Results?

A number of authors and reports have looked at the issue of a "results culture"—an evaluative culture, a culture of inquiry—what it is and how to get there. Based on this literature, an organization with a strong culture of results:

- engages in self-reflection and self-examination:
  - o deliberately seeks evidence on what it is achieving (Botcheva, White and Huffman, 2002; Government Accounting Office, 2003; Smutylo, 2005)
  - o uses results information to challenge and support what it is doing (Hernandez and Visher, 2001)
  - o values candor, challenge and genuine dialogue (David, 2002)
- engages in results-based learning:
  - o makes time to learn (David, 2002)
  - o learns from mistakes and weak performance (Barrados and Mayne, 2003; Goh, 2001)
  - o encourages knowledge transfer (David, 2002; Goh, 2001; Hernandez and Visher, 2001)
- encourages experimentation and change:
  - o supports deliberate risk taking (Pal and Teplova, 2003)
  - o seeks out new ways of doing business (Government Accounting Office, 2003; Goh, 2001; Smutylo, 2005)

Thus, a weaker culture of results might, for example,

- gather results information, but limit its use mainly to reporting,
- acknowledge the need to learn, but not provide the time or structured occasions to learn,
- undergo change only with great effort,
- claims it is results-focused, but discourages challenge and questioning the status quo,
- talk about the importance of results, but frown on risk taking and mistakes, and/or
- talk about the importance of results, but value following rules and procedures and delivering outputs.

Fostering a Culture of Results

Fostering a culture of results is a significant challenge for an organization (Kim, 2002). There are a number of factors that can be identified in building such a "culture of inquiry" (Auditor General of Canada, 2000a; Mayne, 2007c; OECD-DAC, 2006):

- Demonstrated senior management leadership and commitment
- Informed demand for results information
- Supportive organizational systems, practices and procedures
- A results-oriented accountability regime
- A capacity to learn and adapt
- Results measurement and results management capacity

In each of these areas, based on the literature and identified good practices, there are a number of approaches that can be used.

Demonstrated senior management leadership and commitment. Many discussions of implementing results management identify strong senior leadership as essential (Auditor General of Canada, 1997; Binnendijk, 2001; Government Accounting Office, 2002; Perrin, 2006). Providing visible and accountable leadership through such actions as overseeing and supporting the results management regime—identifying and supporting results management champions, walking the talk—providing consistent leadership in results management, challenging the theories of change behind programs and the evidence on past performance, and demonstrating the benefits of results management are key to establishing a culture of results.

Informed demand for results information. A culture of results can be greatly enhanced if managers at all levels, and especially senior levels, consistently and routinely ask for results information in planning, implementing and reviewing contexts (Mackay, 2006; OECD, 2005; OECD-DAC, 2006). That is, a significant role for managers, especially senior managers, in fostering and supporting results management is to routinely ask questions about results and the assumptions behind programs and interventions when reviewing, assessing and making decisions on plans, operations and reports. Knowing that such questions will be forthcoming ensures that those carrying out these tasks will pay attention to results. Asking the questions will help ensure that relevant result information is available when it is needed, and that assumptions about theories of change are routinely challenged. In this way, results information—evidence on what is working, what is not and why—becomes a routine and natural part of managing the organization. Table 6 suggests the nature of those questions.

Supportive organizational systems, practices and procedures. Having the right formal and informal incentives in place is essential to fostering a culture of results. For example, rewarding good managing for results—such as undertaking evaluation, taking informed risk and experimentation, using results information to inform decision making, hold-

Table 6
Results Questions for Management

Results-Based Planning
- What immediate and longer-term results are you trying to achieve?
- How do the intended results align with other priorities of the organization and partners?
- Why do you expect the intended results to be achieved?
- How solid is the theory of change? What evidence is there supporting the assumptions behind how the program is supposed to work? How good is the evidence?
- Who is accountable for what? What are the roles and responsibilities for results of those involved?
- What risks are there to attaining the expected results and how are they to be managed?
- Is the budget commensurate with the expected results?
- What targets for performance have been set?
- What monitoring and evaluation will be undertaken?
- What reporting on results will be done?

Monitored Implementation
- What has been accomplished? What evidence is there that the results you were expecting were achieved? On a day-to-day basis and periodically?
- How do you know your program made a contribution to the observed results?
- Has there been significant variation in the planned budget outlays?
- How well are the risks being managed?

Results-Based Learning
- What have you learned from this past experience with respect to delivery approach, data collection, and the theory of change and its underlying assumptions?
- Has your monitoring and evaluation strategy been modified?

Accounting for Performance
- What is the performance story? What can you credibly report about what has been accomplished?
- How solid is the evidence?

ing learning events, and sharing information on results—demonstrates that the organization does indeed value inquiry and reflection. Wholey (1983) provides a list of incentives to support results management. Osborne (2001) discusses incentives in public sector organizations that are implementing results management. Osborne, Swiss (2005), Wholey (1983) and others all stress the importance of considering non-financial incentives in positively motivating people.

And managers seeking to achieve outcomes need to be able to adjust their operations as they learn what is working and what is not (Binnendijk, 2001; Government Accounting Office, 2002; Norman, 2002). Managing only for planned outputs does not foster a culture of inquiry about what are the impacts of delivering those outputs. Managers need to be provided with the flexibility to adjust their operations in light of the evidence available.

A results-oriented accountability regime. If managers are simply accountable for following procedures and delivering planned outputs, there is little incentive to seek evidence on the outcomes being achieved. And if managers were to be literally accountable for achieving outcomes—which they don't control—they will seek to ensure that accountability is only for outputs. There is a need to adapt the accountability regime to include the idea of influencing outcomes, of being accountability for and rewarding good managing for outcomes (Auditor General of Canada, 2002; Baehler, 2003; Perrin, 2006). Thus, for example, if targets and other expectations have not been met, the most important question should be what has been learned as a result and what will change in the future? On the other hand, managers need to remain accountable for day-to-day delivery. I argue elsewhere (Mayne, 2007a) that accountability for outcomes should consist of:

- Providing information on the extent to which the expected and other outputs and outcomes were attained, and at what cost;
- Demonstrating the contribution made by the activities and outputs of the program to the outcomes;
- Demonstrating the learning and change that have resulted; and
- Providing assurance that the means used were sound and proper.

A capacity to learn and adapt. Learning from empirical evidence and analysis on past performance is what a results culture is all about. Deliberate efforts are needed to build a capacity for and acceptance of learning in an organization. Creating institutionalized learning events (Barrados and Mayne, 2003; Moynihan, 2005), providing group learning opportunities (David, 2002), supportive information sharing and communication structures (Cousins, Goh, Clark and Lee, 2004), making the time to learn and providing adequate resources to do so (David, 2002), seeing mistakes as opportunities to learn (Barrados and Mayne, 2003; Michael, 1993) and focusing on best practices (Pal and Teplova, 2003) are all ways to help foster a culture of learning. Also, evaluative feedback can be used as an approach to facilitate learning between groups of professionals with very different values and orientation (Launsø, Rieper and Rieper, 2007).

In my view, most useful is the idea of institutionalized learning events, which in particular could enhance enlightenment learning in an organization. A learning event could be structured around a current issue of concern where (all) the available information and evidence is brought together in a digestible format for an informed discussion by the interested parties of the issue and what the available evidence implies for future actions. Nutley, Walter and Davies (2007) stress the importance of using "specific deliberative processes" (p. 311) to enhance the use of evidence. They argue that "the interplay between supply and demand of information is what really matters" (p. 311), just what should occur at a learning event. Similarly, Van der Knapp (1995) argues the need for "constructive argumentation" as a way to hence policy-oriented leaning.

Results measurement and results management capacity. Building a culture of results in an organization does require a capacity to be able to articulate and measure results, and a capacity to understand how results information can be used to help managers manage (OECD-DAC, 2006; Perrin, 2002, 2006). Some level of in-house professional results management support is usually required to assist managers and staff (Binnendijk, 2001; Perrin, 2006; Ramage and Armstrong, 2005). Senior and middle managers need to know and understand what result-based management is all about, and how they need to support results management. This capacity can be enhanced through training, using peer champions, and providing senior managers with the kinds of results question they can be routinely asking.

Developing a culture of results in an organization will not happen through good intentions and osmosis. Many organizations face this same challenge. It requires deliberate efforts by the organization and especially its senior managers to encourage and support such a culture. It needs to be clear to managers and staff that results information and evidence is valued and expected to be a regular part of planning, budgeting, implementation, and review.

Implications for Implementing Results Management: Valuing Evidence

This chapter has been examining the issue of institutionalizing results management in an organization, particularly in light of the evidence-movement and what insights it might provide. I began by arguing that results management should be seen as part of the evidence movement.

The problems of getting results information used in decision making and learning are the same as the challenges of getting empirical research evidence accepted as a necessary part of good public management. And by "use" it is widely accepted that results information is meant to inform decisions not determine them. Thus a more accurate term might be *results-informed management*, rather than the more popular results-based management.

Early in the chapter a broad definition of results management was given, but it must be admitted that many, including many results management practitioners, have a narrower concept in mind, namely using data from performance indicators to help managers manage better. This version of results-based management might be called *indicator-based management*.

I have suggested that results management should have a much broader perspective, one that encourages a range of evidence be gathered and used to assist managers. Many have noted the limitations of only relying on indicators (Perrin, 1998; OECD, 2005; Smith, 1995). As the evidence movement literature points out, there is a broad range of types of evidence that results management could make use of:

Evidence from monitoring:

- data gathered from administrative files and resulting statistics,
- data gathered through surveys and direct observation and resulting statistics,

Evidence from studies

- findings from experimental and quasi-experimental designs,
- findings from evaluations, research and other performance studies
- analysis of expert opinions

Evidence from synthesis reviews

- meta-evaluations
- reviews and synthesis of existing research such as done through the Cochrane Centre, the Campbell Collaboration and the Evidence for Policy and Practice Information and Co-ordination Centre.

From this perspective, results management is part of the evidence movement, at least that form of the evidence movement that encourages multiple types of evidence (Learmonth and Harding, 2006; Nutley,

Walter and Davies, 2003, 2007) and not just randomized controlled trials as legitimate evidence. Results management should entail seeking out much richer data and information than what can be obtained through indicator data. This would include evaluations, related research and performance studies, and, building on the evidence movement, synthesis reviews. Moynihan (2005: 203) notes that "most results-based reforms target narrow process improvement (single-loop learning) rather than a broad understanding of policy choices and effectiveness (double-loop learning)."

A key question raised in the evidence-based policy literature is what constitutes legitimate evidence. Unless undertaken dogmatically, this is a healthy debate about the quality of evidence. The quality of evidence in results management has not received adequate attention. Swartz and Mayne (2005) note the differences in approaches to quality among performance audit, evaluation and performance measurement, concluding that performance measurement—the main evidence used by "traditional" results management—has paid considerably less attention to quality issues than has the other two forms of measuring performance of programs. Managers need to be aware of and understand the limitations of the results information and evidence they receive. Seeing results management as part of the evidence-based policy and practice debate would usefully raise the issue of quality and what good evidence entails in results management.

Through comparing traditional management with results management, what is needed from managers in results management was identified: understanding the theory of change behind their programs, measuring results, and using results information to inform decisions on planning, operations and program redesign. Determining and working with theories of change and using results evidence to inform decisions were seen as particularly challenging. It was also noted that results management does not mean a wholesale change from traditional management. On a day-to-day basis, managers continue to need to pay attention to delivery: the inputs being used and the outputs delivered.

But results management, and in particular managing for outcomes, should expect managers to seek to understand why the results observed have occurred and what contribution to the results has their program made. Managers should understand and seek to further clarify and verify the theory of change underlying their programs. Many theories of change are contested, and a focus on them would help encourage informed debate about why programs are working or not, and whether or not other influencing factors are more responsible than the program for any observed outcomes.

In examining a theory of change for using results information/evidence, we saw that to get routine use of results information a number of factors need to be in place. The information has to be "right"—relevant, timely, credible and understandable. However, the biggest challenge to getting results information routinely used in organizations is the need to have interest in results information commonplace; a results-oriented culture is often absent. Organizations often do not see evidence on past experience as an essential part of routine good management.

A theory of change for the routine use of results information would then be:

> IF there is a results-oriented culture in the organization, AND
> IF the "right" results information is available, THEN
> The results information will routinely be used to inform decisions.

This theory of change may not appear too noteworthy, but I think it is. It is saying that you cannot just focus on getting the "right" results information—which is still a challenge—but even more important is fostering and nurturing a results-oriented culture in the organization. Hernandez and Visher (2001) conclude similarly in an assessment of implementing results measurement in a number of not-for-profit organizations. Without that culture and ongoing efforts to enhance and support it, at most, only sporadic use of results information will occur.

The issue of building a results-oriented culture in an organization, despite being widely recognized and a key challenge, has received rather less attention than the issue of getting the "right" information, where there is vast literature and numerous guide and discussions of good practices in evaluating and monitoring results. By stressing the essential role of culture and outlining some the requirements and possible approaches to fostering such a culture, I hope that in the future there will be much more good practices in the literature, such as those discussed by Hernandez and Visher (2001), in developing results-based cultures in organizations.

In the End: Making Time to Learn

The bottom line is that too often, seeking and considering evidence is not part of management. Managers often don't see themselves as having enough time to manage for results. Their days are fully occupied in planning, keeping the program on course, fighting fires and reporting. At best, they may be able to react to evidence gathered from monitoring that identifies problems arising such as planned outputs not being delivered. The more in depth reflection on the theory of change of the program and the extent to which the program is contributing to the desired outcomes may be seen as a luxury few can afford.

If this indeed the case, managing for outcomes will remain weak. I would argue that a large part of the problem is that "learning" is not institutionalized as a necessary part of managing in the same way planning is. Time is always made for planning. Time also needs to be made for learning, and learning events need to be institutionalized, in the same way planning events are. And strategic planning events may not be the best time to learn (Moynihan, 2005), since they have other objectives in mind-goal setting not process change, and may not be as inclusive as needed for organizational learning. Britton (2005) discusses a variety of ways organizations can make "space" for learning.

At scheduled learning events, the latest available empirical and other evidence on a program or issue could be considered and debated. Implications for future programming could be discussed and gaps in evidence could be identified. Learning events provide the time and space to digest the richer information than is available from just monitoring data. A culture of results would reinforce such practice as normal good management. These ideas, which have come from the world of results management, may be something that the evidence movement can learn from.

References

Auditor General of Canada (1997). *Moving towards managing for results.* Report of the Auditor General of Canada to the House of Commons, October. Ottawa. http://www.oag-bvg.gc.ca/domino/reports.nsf/html/ch9711e.html (Retrieved September 3, 2007)

Auditor General of Canada (2000a). *Managing Departments for Results and Managing for Horizontal Issues for Results.* Report of the Auditor General of Canada to the House of Commons, October. Ottawa.http://www.oag-bvg.gc.ca/domino/reports.nsf/html/0020ce.html/$file/0020ce.pdf (Retrieved September 3, 2007)

Auditor General of Canada (2000b). *Reporting performance to parliament: Progress too slow.* Report of the Auditor General of Canada to the House of Commons, December. Ottawa. http://www.oag-bvg.gc.ca/domino/reports.nsf/html/0019ce.html/$file/0019ce.pdf (Retrieved September 3, 2007)

Auditor General of Canada (2002). *Modernizing accountability in the public sector. Report of the Auditor General of Canada to the House of Commons, Chapter 9.* Ottawa. http://www.oag-bvg.gc.ca/domino/reports.nsf/html/20021209ce.html/$file/20021209ce.pdf (Retrieved September 3, 2007)

Australian National Audit Office (2007). *Application of the outcomes and iutputs framework.* Canberra. Audit Report No.23 2006–07. http://www.anao.gov.au/uploads/documents/2006-07_Audit_Report_23.pdf (Retrieved September 3, 2007)

Baehler, K. (2003). "'Managing for outcomes': Accountability and thrust." *Australian Journal of Public Administration* 62,4: 23-34.

Barrados, M. and J. Mayne (2003). "Can public sector organizations learn?" *OECD Journal on Budgeting* 3,3: 87-103.

Binnendijk, A. (2001). *Results-based management in the development cooperation agencies: A review of experience.* Background Report, DAC OECD Working Party on Aid Evaluation. Paris. http://www.oecd.org/dataoecd/17/1/1886527.pdf (Retrieved 3 September 2007).

Boston, J., J. Martin, J. Pallot and P. Walsh (eds.) (1991). *Reshaping the state: New Zealand's bureaucratic revolution.* Auckland: Oxford University Press.

Botcheva, L., C.R. White and L.C. Huffman (2002). "Learning culture and outcomes Measurement practices in community agencies." *American Journal of Evaluation* 23,4: 421-434.

Britton, B. (2005). *Organisational learning in NGOs: Creating the motive, means and opportunity*, Praxis Report No. 3: The International NGO training and Research Centre. Available at http://www.intrac.org/pages/PraxisPaper3.html (accessed 20 Feb 2008).

Caracelli, V. J. and H. Preskill (eds.) (2000). *The expanding scope of evaluation use.* New Directions for Evaluation, No. 88. San Francisco: Jossey-Bass.

Chalmers, I. (2005). "If evidence-informed policy works in practice, does it matter if it doesn't work in theory?" *Evidence & Policy: A Journal of Research, Debate and Practice*, 1,2: 227-242.

Cousins, B., S. Goh, S. Clark and L. Lee (2004). "Integrating evaluative inquiry into the organizational culture: A review and synthesis of the knowledge base." *Canadian Journal of Program Evaluation*, 19,2: 99-141.

Curristine, T. (2005). "Government performance: Lessons and challenges." *OECD Journal on Budgeting* 5,1: 127-151.

David, T. (2002). *Becoming a learning organization*: Marguerite Casey Foundation. http://www.tdavid.net/pdf/becoming_learning.pdf (Retrieved 3 September 2007)

Davies, H.T.O., S.M. Nutley and P.C. Smith (1999). "Editorial: What works? The role of evidence in public sector policy and practice." *Public Money and Management* 19,1: 3-5.

Diamond, J. (2005). *Establishing a performance management framework for government.* IMF Working Paper, International Monetary Fund. http://www.imf.org/external/pubs/ft/wp/2005/wp0550.pdf (Retrieved September 3, 2007)

Feinstein, O.N. (2002). "Use of evaluations and the evaluation of their use." *Evaluation* 8,4: 433-439.

Furubo, J.-E. (2006). "Why evaluations sometimes can't be used—and why they shouldn't. In R.C. Rist and N. Stame (eds.), *From studies to streams: Managing evaluative streams*, New Brunswick, NJ, Transaction Publishers: 147-166.

Goh, S. (2001). "The learning organization: An empirical test of a normative perspective." *International Journal of Organizational Theory and Behaviour*, 4(3&4): 329-355.

Goldsworthy, D. (1991). *Setting up next steps.* London: HMSO.

Government Accounting Office (2002). *Results-oriented cultures: Insights for U.S. agencies from other countries' performance management initiatives.* Washington: U.S. General Accounting Office. http://www.gao.gov/new.items/d02862.pdf (Retrieved September 3, 2007)

Government Accounting Office (2003). *An evaluation culture and collaborative partnerships Help Build Agency Capacity. Program Evaluation.* Washington, DC. http://www.gao.gov/new.items/d03454.pdf (Retrieved 3 September 2007)

Hernandez, G. and M. Visher (2001). *Creating a culture of inquiry: Changing methods—and minds—on the use of evaluation in nonprofit organizations.* The James Irving Foundation.

Johnson, C. and C. Talbot (2007). The UK parliament and performance. *International Review of Administrative Sciences*, 73(1):113-131.

Joyce, P.G. (2005). "Linking performance and budgeting: Opportunities in the federal budget process." In. J.M. Kamensky and A. Morales (eds.), *Managing for Results 2005,* Oxford: Rowman and Littlefield.

Kamensky, J.M. and A. Morales (eds.) (2005). *Managing for results 2005.* The IBM Center for The Business of Government. Oxford: Rowman and Littlefield.

Kim, P.S. (2002). *Cultural change in government: Promoting a high-performance culture A Review of ten years of modernisation: The HRM Perspective.* Paris: OECD http://www.olis.oecd.org/olis/2002doc.nsf/LinkTo/PUMA-HRM(2002)11

Kogan, M. (1999). "The impact of research on policy." In F. Coffield (ed.), *Research and policy in lifelong learning,* Bristol: Policy Press.

Launsø, L., J. Rieper and O. Reiper (2007). "Evaluative feedback as a contribution to learning between groups of professionals." *Evaluation* 13,1: 306-322.

Learmont, M. and N. Harding (2006). "Evidence-based management: The very idea." *Public Administration* 8,2: 245-266.

Leeuw, F. (2006). "Managing evaluations in the Netherlands and types of knowledge." In R.C. Rist and N. Stame (eds.), *From studies to streams: Managing rvaluative systems,* New Brunswick, NJ, Transaction Publishers: 81-96.

Leeuw, F.L., R.C. Rist and R.C. Sonnichsen (eds.) (1994). *Can governments learn? Comparative Perspectives on Evaluation and Organizational Learning,* New Brunswick, NJ: Transaction.

Leeuw, F., L.M. v. d. Knaap and S. Bogaerts (2007). "Reducing the knowledge-practice gap: An example of an evaluation synthesis in criminal policy." *Public Money & Management* 27,4: 245-250.

Leviton, L.C. (2003). "Evaluation use: Advances, challenges and applications." *American Journal of Evaluation* 24,4: 525-535.

Leviton, L.C. and E.F.X. Hughes (1981). "Research on the utilization of evaluation: A review and synthesis." *Evaluation Review* 5,4: 525-548.

Lindquist, E. (2001). *Discerning policy influence: Framework for a strategic evaluation of IDRC-supported research.* University of Victoria. http://www.idrc.ca/uploads/user-S/10359907080discerning_policy.pdf (Retrieved September 3, 2007)

Mackay, K. (2006). *Institutionalization of monitoring and evaluation systems to improve public sector management EDC Working Paper Series- No. 15.* Washington, DC: Independent Evaluation Group, World Bank. http://www.worldbank.org/ieg/ecd/ (Retrieved September 3, 2007)

Mark, M. and G. Henry (2004). "The mechanisms and outcomes of evaluation influence." *Evaluation* 10,1: 35-57.

Marra, M. (2004). "The contribution of evaluation to socialization and externalization of tacit knowledge: The case of the World Bank." *Evaluation* 10,3: 263-283.

Mayne, J. (2001). "Addressing attribution through contribution analysis: Using performance measures sensibly." *Canadian Journal of Program Evaluation* 16,1: 1-24.

Mayne, J. (2007a). "Evaluation for accountability: Reality or myth?" In M.-L. Bemelmans-Videc, J. Lonsdale and B. Perrin (eds.), *Making accountability work: Dilemmas for evaluation and for audit.* New Brunswick, NJ, Transaction Publishers.

Mayne, J. (2007b). "Challenges and lessons in implementing results-based management." *Evaluation* 13,1: 89-107.

Mayne, J. (2007c). *Best practices in results-based management: A review of experience— A report for the United Nations Secretariat.* New York: UN Secretariat.

Mayne, J. and E. Zapico-Goni (eds.) (1997). *Monitoring performance in the public sector: Future directions from international experience.* New Brunswick, NJ: Transaction Publishers.

Mayne, J. and R. Rist (2006). "Studies are not enough: The necessary transformation of evaluation." *Canadian Journal of Program Evaluation.*

Michael, D. (1993). "Governing by learning: boundaries, myths and metaphors." *Futures* January/February:81-89.

Moynihan, D.P. (2005). "Goal-based learning and the future of performance management." *Public Administration Review* 65,2: 203-216.

Moynihan, D.P. (2006). "Managing for results in state government: Evaluating a decade of reform." *Public Administration Review* 66,1:77-89.

Norman, R. (2002). "Managing through measurement or meaning? Lessons from experience with New Zealand's public sector performance management systems." *International Review of Administrative Sciences*, 68:619-628.

Nutley, S., I. Walter and H.T.O. Davies. (2003). "From knowing to doing: A framework for understanding the evidence-into-practice agenda." *Evaluation* 9,2: 125-148.

Nutley, S.M., I. Walter and H.T.O. Davies (2007). *Using evidence: How research can inform public services.* Bristol: The Policy Press.

OECD (2005). *Modernizing government: The way forward.* Paris, OECD.

OECD-DAC (2006). *Managing for development results, Principles in action: Sourcebook on emerging good practices (Final—March 2006).* Paris: OECD-Development Assistance Committee. http://www.mfdr.org/Sourcebook/1stSourcebook.html (Retrieved September 3, 2007)

Osborne, D. (2001). *Paying for results.* Government Executive, February: 61-67.

Pal, L. A. and T. Teplova (2003). *Rubik's cube? Aligning organizational culture, performance measurement, and horizontal management.* Ottawa: Carleton University. http://www.ppx.ca/Research/PPX-Research%20-%20Pal-Teplova%2005-15-03[1].pdf (Retrieved September 3, 2007)

Pawson, R. (2002a). "Evidence-based policy: In search of a method." *Evaluation* 8,2: 157-181.

Pawson, R. (2002b). "Evidence-based policy: The promise of 'realist synthesis'." *Evaluation* 8,3: 340-358.

Perrin, B. (1998). "Effective use and misuse of performance measurement." *American Journal of Evaluation* 19,3: 367-379.

Perrin, B. (2002). *Implementing the vision: Addressing challenges to results-focussed management and budgeting.* Paris: OECD.

Perrin, B. (2006). *Moving from outputs to outcomes: Practical advice from governments around the world.* IBM Centre for The Business of Government and the World Bank. Washington, DC.

Pollitt, C. (2006). "Performance information for democracy: The missing link?" *Evaluation*, 12,1: 38-55.

Pollitt, C. and G. Bouckaert (2000). *Public management reform: A comparative analysis.* New York, Oxford University Press.

Ramage, P. and A. Armstrong (2005). "Measuring success: Factors impacting on the implementation and use of performance measurement within Victoria's human services agencies." *Evaluation Journal of Australasia*, 5(new series),2: 5-17.

Rist, R. and N. Stame (eds.) (2006). *From studies to streams: Managing evaluative systems*: Transaction Publishers.

Sanderson, I. (2002). "Evaluation, policy learning and evidence-based policy making." *Public Administration* 80,1: 1-22.

Sanderson, I. (2004). "Getting evidence into practice: Perspectives on rationality." *Evaluation* 10,3: 366-379.

Schwartz, R. and J. Mayne (2005). Does quality matter? Who cares about the quality of evaluative information? In *Quality matters: Seeking confidence in evaluation, auditing and performance reporting*. R. Schwartz and J. Mayne, Eds. New Brunswick, Transaction Publishers.

Smith, P. (1995). "On the unintended consequences of publishing performance data in the public sector." *International Journal of Public Administration* 18,2&3: 277-310.

Smutylo, T. (2005). *Building an rvaluative culture*. International Program for Development Evaluation Training. World Bank and Carleton University. Ottawa.

Sridharan, S. (2003). "Introduction to special section on 'What is a useful evaluation?'" *American Journal of Evaluation* 24,4: 483-487.

Swiss, J. (2005). "A framework for assessing incentives in results-based management." *Public Administration Review* 65,5: 592-602.

ter Bogt, H. (2004). "Politicians in search of performance information? Survey research on Dutch aldermen's use of performance information." *Financial Accountability and Management* 20,3: 221-252.

The Council for Excellence in Government (2008). *Coalition for evidence-based policy*. Washington, DC. Available at http://www.excelgov.org/admin/FormManager/filesuploading/Coalition_purpose_agenda_3_06.pdf.

Thomas, G. (2004): "Introduction: evidence and practice." In G. Thomas and R. Pring (eds.), *Evidence-based practice in education*, New York: Open University Press.

Torres, R.T., H.S. Preskill and M.E. Piontek (2005). *Evaluation strategies for communicating and reporting: Enhancing learning in organizations*. Thousand Islands: Sage.

Valovirta, V. (2002). "Evaluation utilization as argumentation." *Evaluation*, 8,1: 60-80.

van der Knaap, P. (1995). "Policy evaluation and learning: feedback, enlightenment or argumentation?" *Evaluation* 1,2: 189-216.

Vedung, E. (1997). *Public policy and program evaluation*. New Brunswick: Transaction Publishers.

Walter, I., S. Nutley and H. Davies (2005). "What works to promote evidence-based practice? A cross-sector review." *Evidence & Policy: A Journal of Research, Debate and Practice* 1,3: 335-364.

Weiss, C.H. (1998). "Have we learned anything new about the use of evaluation?" *American Journal of Evaluation* 19,1: 21-33.

Wholey, J.S. (1983). *Evaluation and effective public management*. Boston: Little, Brown and Co.

Williams, D. (2003). "Measuring government in the early twentieth century." *Public Administration Review* 63,6: 643-659.

World Bank-OED (2005). *2004 Annual Report on Operations Evaluation*. Washington, DC: World Bank-Operations Evaluation Department. http://www.worldbank.org/oed/popular_corporate.html (Retrieved 3 September 2007)

8

Evidence and Politicians: Can the Two Co-exist?

Richard Boyle

Introduction

As has been well documented, there has been a rapid growth since the 1980s in the amount of evidence produced by public bodies (see for example Mayne and Zapico-Goni, 1997; OECD, 2005; Johnson and Talbot, 2007). Evidence is described here as including not only information from randomized trails and systematic reviews, but also evaluative information more generally, particularly information arising from evaluations and performance reporting. In some cases, this growth has occurred because of legislative drivers, such as the Government Performance and Results Act of 1993 in the United States and the Public Service Management Act, 1997 in Ireland. In other cases, such as the United Kingdom, Netherlands and Spain, general modernization initiatives for the public service have been the spur. Whatever the impetus, one of the aims of this increase in evidence is to better inform politicians of the value and impact of public expenditure.

A central tenet of the evidence movement is that enhanced information on the performance of public services, by initiatives such as above, and in particular through greater use of randomized trials and systematic reviews, will lead to more informed political decision making. But are politicians using this evidence? An OECD (2007) overview found that of 32 countries surveyed, in only six cases was it reported that the legislature is involved in monitoring progress against performance measures. Similarly, again in only six countries was it reported that the legislature applies performance results in resource allocation and/or program or policy decisions. Pollitt (2006) in a review of the use of performance information by politicians and citizens further notes that:

> Thus far the picture conveyed by the (fragmentary) academic literature looks gloomy. Grand statements about the importance of performance information for democracy sit alongside extensive if patchy evidence that ministers, legislators and citizens rarely make use of the volumes of performance information now thrust upon them.

This picture is further corroborated by Johnson and Talbot (2007), who undertook a review of UK parliamentary select committee scrutiny of Public Service Agreements (PSAs, high level output, and outcome focused targets). They found that departmental select committees scrutinize only a small proportion of PSA targets, and that those that are scrutinized are often only touched on lightly in the course of discussion and debate. They conclude:

> Taken together these analyses can only lead to the conclusion the PSAs are not scrutinized broadly (only around 1/5th get any coverage) and of those that do get mentioned they are mostly tangential or a relatively small part of reports. This certainly does not add up to scrutiny of "output or outcome-based government" if PSAs are the main vehicle for reporting achievement.

Nor is this limited use of evaluative evidence on performance by politicians restricted to Anglo Saxon experience. Bogt (2004) carried out research amongst Dutch local government politicians (aldermen), and found that the majority of aldermen see limited value in the output-oriented performance information available to them. They prefer meetings and consultations with civil servants as a means of communicating and receiving evidence: "In general, all aldermen seemed to prefer rich, verbal information to sources of written information probably because they work in a relatively complex and uncertain political environment" (Bogt, 2004, p. 241).

There is a further issue, that politicians use evidence, but use it selectively to support pre-existing positions. Quinn (2008) notes:

> Despite the focus frequently placed on "evidence based" policy making, political priorities and the political process are acknowledged to be the determining factors for the influence which a particular research study garners. Parliamentary questions (PQs) which ask the Minister if he will accept the recommendations/ views of a particular agency report typically generate two types of replies: if the research is consistent with and supports the current policy agenda it will be welcomed as an important "evidence based" element in justifying the policy decision: however, where the research opposes or fails to support the established view, it will be welcomed as an important "contribution" to the overall debate, but may ultimately be ignored. An example of this can be seen in PQs concerned with the extensive research on poverty measurement and the "risk of poverty" rate in Ireland. This research, accompanied by demands of state agencies and others to develop targets in this area, has been ef-

fectively dismissed as government continues to focus on poverty in a more absolute or deprivation model. Research, irrespective of its quality and strength of conclusions, will not change policy if there is no official support (p. 92).

Thus, we have a situation where politicians' use of evidence appears to be limited in both quantity and quality. Often, politicians will not make use of available evidence, and when they do, they may choose to use the evidence selectively. Why should this be so? Do initiatives such as systematic reviews and a greater emphasis on empirical data, as mentioned in the Introduction to this book, change the picture? It is questions such as these that this chapter addresses.

Politicians—A Homogenous Group?

Before looking at how politicians use or don't use evidence, it is useful to reflect on the identification of politicians as a stakeholder group with regard to evidence. In discussing politicians and their use of evidence as exemplified by their use of evaluative information, it is common to see references to politicians as if they are a homogenous group, with common needs and responses to evaluative information. However, in assessing the relationship between politicians and evidence, it is helpful to go beyond this simple view and to bear in mind different categorizations of politicians and how they might interact with evaluative information about the performance of public services.

At the national level, three broad groupings of politicians can be identified for the purposes of this exercise: ministers, members of parliamentary committees and individual parliamentarians. At the local level, equivalent categorizations are possible.[1] Ministers, as members of the executive, are supported by staff of various sizes and might be expected to engage with evidence on performance in the course of policy formulation and implementation monitoring. Parliamentary committees are intended to hold ministers to account, an important part of the process of scrutiny and accountability of the executive to parliament. In this scrutiny role, committee members engagement with evidence on performance might be expected to both test the validity of evidence presented by the executive, and also to produce their own evidence on pertinent issues. Committee members tend to have limited staff supports. As individual parliamentarians, politicians often act primarily as representatives of their constituencies. As such, they have relatively little support from staff and their engagement with evidence on performance might be expected to be more ad hoc, with the media and the views of constituents playing an important role in formulating opinions and stances on issues (Likierman, 1988; Lomas, 2007).

Parliamentary committees pose a particularly interesting grouping of politicians from the perspective of engagement of politicians with evidence. Particular attention is given in this paper to the interaction between parliamentary committees and evidence. In theory, committees are intended to enhance the scrutiny of the executive. Committee membership allows members to build up expertise in certain areas that they can bring to bear in their scrutiny role. But in practice the picture is not necessarily so positive. MacCarthaigh (2005) notes that with regard to Irish experience with committee work, in terms of better scrutiny of the executive and its bureaucracy, the parliamentary committee experiment to date leaves much to be desired. Politicians claim that they do not have the time to engage in detailed scrutiny of the administration and, tellingly, they question the electoral benefits in return for such work. MacCarthaigh further states that: "Committee chairpersons tend to pursue matters that are of interest to committee membership, and an examination of recent committee reports shows that scrutiny of departmental financial estimates is practically non-existent" (p. 142).

A Theory of Change for Politicians' Use of Evidence

Mayne (this volume) outlines a theory of change for using information on results. His main focus is on the use of information within an organization by senior managers. A similar approach is taken here in developing a theory of change for the use of evidence by politicians. The literature suggests that politicians as end users are making little use of evidence available to them in the form of evaluative information on public service performance. What is the theory of change that informs thinking about how politicians engage with evidence? Table 1 sets out a theory of change for politicians' use of evidence. The theory defined here broadly follows the logic of reviews of how research is intended to impact on policy, ranging from the production and receipt of evidence, through the extent to which evidence is considered, to the extent evidence is seen as significant, increases knowledge, and changes attitudes and behaviors (Nutley, 2003, p. 8).

Evidence is Produced in an Accessible and Timely Manner

It is well known that how evidence is presented influences reactions to and use of evidence by those receiving it. Politicians are no different from others in this regard. Brun and Siegel (2006), in a survey of Swiss experience that included cantonal government politicians (full time professionals and part of the executive, supported by staff) and

Table 1
Theory of Change for Politicians' Use of Evidence

Assumption 1	Evidence is produced in an accessible and timely manner
Assumption 2	Politicians receive the evidence
Assumption 3	Politicians can interpret and make sense of the evidence
Assumption 4	Politicians view the evidence as being of significance
Assumption 5	Politicians use the evidence in discussions and debates surrounding decisions

cantonal parliamentarians (people's representatives, with little staff support) found that there was no significant difference between the two groups surveyed regarding their views on how performance information should be delivered. The following elements of presentation were mentioned as being helpful to politicians when presenting performance information to them:

- A standardized comparable layout is generally desired
- The visualization of the content with the aid of tables and figures
- Presentation by means of indicators, as well as commentary
- A "dashboard" which presents the information clearly, concentrating on essentials, and providing data on the degree of goal achievement (Brun and Siegel, 2006, p.491).

With regard to evidence being produced in an accessible and timely manner, MacCarthaigh (2005) illustrates, in the case of parliamentary committees, the key role that professional staff support can play in ensuring that politicians get the evidence they require. A sub committee of the Public Accounts Committee was established to conduct an inquiry into tax evasion associated with a particular tax scheme.[2] The Office of the Comptroller and Auditor General prepared the evidence base for the inquiry prior to the sub committee meeting. This allowed the sub committee to take evidence from all parties in less than six weeks. Mac-Carthaigh notes that "this was in sharp contrast to tribunals of inquiry … where much of the time was spent collecting and searching for evidence before hearings could get under way" (p. 181).

Politicians Receive the Evidence

It should not be assumed that just because evidence is produced, the intended users of that evidence will receive it. Politicians are extremely busy individuals, with many competing demands on their time. Evaluative evidence on the performance of the public service, just because it is produced, does not necessarily automatically get to the political audience it is intended for.

This point is illustrated by evaluation reports carried out by Irish government departments as part of the Government's Policy and Value for Money Review.[3] Initially, as a way of facilitating political engagement with the findings from the evaluation reports on expenditure programs, the reports were to be laid before the Houses of the Oireachtas (Irish parliament). In the 2002-04 round of reviews, forty four reviews were undertaken and submitted to the Houses of the Oireachtas. But none of these reports were discussed in the Oireachtas. Simply making them available did not mean that they were taken up. To engage political attention, it seems that, amongst other things, there needs to be more structured presentation of the evidence. In the case of Value for Money and Policy Reviews, from 2006 departments must now send copies of the evaluation reports to the clerks of appropriate Oireachtas committees. Parliamentary clerks have been advised to explicitly draw the attention of committee members to the existence of the reports and to encourage debate at select committee meetings on the report findings.

A further illustration of politicians failing to receive evidence is provided by the case of agencies in Ireland charged with producing evidence to inform policy formulation. Quinn (2008) in a study of agencies involved in social policy issues, suggests that considerable time and public money is wasted in producing research—both first and second order evidence—which has no impact. This she puts down in large part to the absence of agencies from the main institutional structures set up to support politicians, and in particular cabinet members, in their policy role, particularly social partnership and cabinet committees. In interviews with agency chief executives, she noted no engagement between agencies and cabinet committee structures, and a lack of awareness of their importance. In particular, senior officials groups that support cabinet committees are key gatekeepers and disseminators of evidence to cabinet members, but were not seen in this light by the agencies. Although less focused on the role of parliament and parliamentary committees, this example illustrates that even within government agencies there can be an assumption that once they have produced evidence it will be taken up by the political system to inform decision making. This is not the case, and knowledge of the structures and processes by which evidence is brought to the attention of politicians is needed if we are to understand and address this issue.

Politicians Can Interpret and Make Sense of the Evidence

Once politicians receive the evidence, they then must interpret and make sense of that evidence. Often, this is most effective when they are supported in the process. In the case of investigation of tax evasion by a sub committee of the Public Accounts Committee in Ireland mentioned above, MacCarthaigh (2005) noted the key role of the Office of the Comptroller and Auditor General. The committee members did not have the technical knowledge necessary to investigate the financial details involved in the inquiry. It is not unusual for parliamentary committees to receive support from specialist staff to help them make sense of the evidence received. Similarly, as mentioned above, cabinet committees are often supported by senior official groups (Quinn, 2008). These senior official groups are important "filters" of evidence for cabinet members.

Other "filters" of evidence to politicians to guide their interpretation and sense making of evidence are what are described by Moynihan (2005a) as "advocates." With regard to parliamentarians' engagement with budget information, Moynihan describes the central role that advocates play in filtering the evidence and guiding politicians in their interpretation and sense-making process:

> Decision makers who receive budget information from agencies lack the time, interest, and capacity to make decisions on resource allocation based on performance information. The budget cycle is too short, and budgeters have enough to do to make budget decisions incrementally, even with all of the heuristics that are associated with this approach.... It is nearly impossible for them to make judgments on what mounds of performance information indicate for a particular functional area. This is true even if only a couple of performance measures are reported per program. There will never be enough information to substitute for the expertise and knowledge to enable substantive judgments to be made, and there will always be too much information for human cognitive processes to deal with.

> Advocates help make performance information understandable. Agency advocates direct attention to particular pieces of information, framing what it means in terms of resource needs and offering decision suggestions for these audiences (Moynihan, 2005a, p. 239).

Moynihan (2005b) further explains that advocates often use what he terms "learning metaphors" as a means of getting evidence understood by politicians. He describes how staff in the Department of Corrections in Vermont used performance information to shape policy outcomes and allow the department to convince others, including politicians, of the legitimacy of these outcomes. To do this, they employed what Moynihan describes as a series of "metaphors," including managing for results, as

a means of persuading politicians and other stakeholders to look positively on evidence that showed the benefits of restorative rather than just
redistributive or rehabilitative justice. Moynihan (2005b, p. 208) states
that: "the credibility of the metaphors comes from their association
with a respected body of knowledge or way of thinking and operating."
Further details are given in Box 1. What is clear is that how evidence is
packaged and presented to politicians influences how they interpret and
make sense of it.

Politicians View the Evidence as Being of Significance

Once politicians have received and made sense of the evidence, they
must engage with it if it is to have meaning to them. It must be seen as
significant enough as to merit their doing something with it. A com-

Box 1
**Using Learning Metaphors to Present Evidence to Politicians on Corrections
Policy in Vermont**

Many politicians and other stakeholders held a traditional view of corrections,
with an emphasis on a punitive approach to incarceration. Corrections staff used
metaphors such as managing for results, business and science to present evidence
that showed an alternative approach based on restorative justice was more likely to
be effective. For example, analysis was used to show that the punitive approach to
incarceration led, overall, to reduced levels of public safety, as those incarcerated
had higher recidivism rates than those placed in alternatives to incarceration, even
after controlling for the risk profile of prisoners. A risk management approach was
proposed based on the severity of the crime and the potential to re-offend. This risk
management approach was portrayed as non-ideological, pragmatic and scientific
in nature. One official noted that "data has allowed us to establish significant credibility with the legislature."

The Department of Corrections had a high degree of success in convincing the state
government and the public that a punitive approach to corrections is wrong, and
that rehabilitation should be the course pursued in certain circumstances. These
arguments ran counter to the prevailing view throughout the country, where pro-
ponents of such arguments have had limited success and have been characterized
as soft on crime. Corrections officials in Vermont were largely able to avoid this
characterization by emphasizing public safety and arguing for increased incarceration for violent offenders.

The department managed to build up strong support from stakeholders with high
political legitimacy, such as victims groups and reparative boards. This support
was seen as crucial in winning over the politicians.

Source: Moynihan, 2005b

monly cited argument concerning the production of evidence is that the evidence production cycle is "out of synch" with the political cycle. Evidence is not vested with significance because it is not seen as timely by political users. The enlightenment model of use, however, suggests other possibilities. Thomas (1985) refers to the relationship of research and policy as a limestone model, where research may drip through into a stagnant lake of overlooked knowledge until eventually some decision maker may appreciate its significance and it may reach political bedrock. But such enlightenment takes time and is uncertain in its application.

Adapting the rational choice model of information for managers developed by Daft and Macintosh (1978) to politicians, it may be assumed that politicians attach significance to simple, lean information, such as numeric information in reports, in low uncertainty situations. But in highly uncertain situations, the work of Daft and Macintosh would suggest that politicians attach significance to rich, flexible and personally obtained information. This point is supported by Bogt (2004) in his survey of Dutch aldermen. He found that of those aldermen who responded to his survey, the majority frequently get their information during informal, verbal consultations with professional top managers. The second most important source of information is formal meetings and consultations with top managers. Bogt also finds that: "Of the available sources of formal, written information, the reports by civil servants and policy notes were used quite frequently. However, much less use was made of budgets, annual reports, and interim reports, which contained formal, written, largely numeric information" (Bogt, 2004, p. 235).

Of course, a key element is whether evidence highlights issues of concern to politicians' constituents and so is relevant to the electability of candidates. This is notable in the political and public discussion on fluoridation of water in Ireland. Here, the evidence from a number of systematic reviews is strongly in favor of fluoridation. But some politicians have attached significance to other sources of research evidence suggesting health dangers associated with fluoridation (see Box 2 for details). These health dangers have been highlighted by environmental lobby groups in particular. Despite the weight and standing of the evidence of the systematic reviews, the less weighty evidence is gaining credence, due in part to the significance attached to it by some politicians in response to public concerns.

Box 2
Water Fluoridation: Alternative Views of the Significance of Evidence

Public mains drinking water in Ireland is artificially fluoridated as a public health measure aimed at preventing tooth decay. An overview of water fluoridation based on systematic reviews and other evidence highlighted the positive benefits, including reduced tooth decay rates (Department of Health and Children, 2002). The only negative effect reported is a mottling of the tooth enamel, a largely treatable cosmetic effect impacting on a small percentage of people. There is no water fluoridation in Northern Ireland, enabling comparative data to be produced. Decay rates were similar on both sides of the border pre fluoridation. Decay rates in Northern Ireland are now about 50 per cent higher than in the Republic of Ireland, although fluoride toothpaste usage seems to be higher in Northern Ireland.

Yet despite this evidence, politicians are being influenced by other single research studies that suggest health problems associated with fluoridation. The Green Party in Ireland has stated it will stop fluoridation of water supplies if elected to government. They cite American studies suggesting a danger to babies and bone cancer risks. A respected scientist and commentator in this area notes that "to take these charges seriously is to champion the odd negative report over the very many studies, including several huge reviews of the literature, that find no evidence of a link between water fluoridation and ill-health." (Reville, 2007). He notes that it is much easier to seed doubt and fear in people's minds than it is to seed reassurance.

Politicians Use the Evidence in Discussion and Debate Surrounding Decisions

The ultimate test of evidence for political decision making is that the evidence is actively used by politicians in their discussions and debates around policy decisions. Often, policy change is as a result of the accumulation of a critical mass of evidence, with findings pointing relatively unambiguously in a particular direction.

This can be exemplified by the introduction of the smoking ban in Ireland. Ireland was the first country in the world to introduce a nation wide ban on smoking in the workplace. For those with knowledge of Irish pub culture, such a step is no small achievement. Barrington (2007) notes that the smoking ban is a good example of research being translated from evidence into action. She states:

A long-standing commitment by Government policymakers to reduce tobacco consumption was in place from as far back as the 1970s. An active, vocal and articulate public health community campaigned for further regulation, which led to the Department of Health and Children and the Health and Safety Authority commissioning a systematic review of research findings on the environmental impact of tobacco smoke. The review highlighted the damage secondary smoking was doing to the health of workers, leading to a Government commitment to prohibit smoking in work places to protect the health of workers.

If the fluoridation issue noted above is an example where systematic reviews do not necessarily end all debate amongst politicians, the smoking ban is an example of where a systematic review was a powerful tool in policy change. This would suggest that it is not just the evidence from systematic reviews themselves that is important when it comes to political use of the evidence, but the context within which the evidence is presented. Going back to the Introduction of this chapter, and the issues raised there, the suggestion is that systematic reviews do not, of themselves, necessarily convince politicians that they represent a higher level of evidence than other evidence sources. It is only when the context within which the systematic review evidence is presented is supportive that systematic review evidence is given particular weight.

Also, it is notable that if politicians fail to show an interest in evidence for decision making, this feeds back to influence the behavior of officials. Moynihan (2005a) shows how, in the case of performance information for budgeting in Alabama, elected officials had become so indifferent to performance information that it was no longer deemed important enough to be published in the budget. Moynihan notes that this was "…a point not lost on agency officials, who complied with legislative requirements by delivering poor quality performance information, if any" (p.232).

Conclusions

There is little new in the news that politicians do not automatically use evidence. The evaluation literature has discussed issues around stakeholder use or non use of evaluation findings for several decades. Similarly research impact studies have identified common reasons why research often does not influence decision makers, including politicians. What this paper tries to do is to put some structure on our understanding of why there is a limited interaction between evidence and politicians. Using a theory of change model derived from a rather linear, rational understanding of how evidence should feed through into political decision making, limitations in practice at different stages in the process have been identified.

In thinking how to improve political engagement in the future, and learning from the experiences outlined, one step is the development of a revised theory of change model that might more accurately reflects how evidence might ultimately influence political decision making. Such a revised model is set out in Table 2. It is put forward not as an answer to the problem of engaging politicians with evidence, but as a prompt for further discussion and articulation of our understanding of the theory behind politicians' engagement with evidence.

The revised model retains some of the elements of the previous model. The changes relate to assumptions two and three, that politicians receive the evidence and that politicians interpret and make sense of the information. Drawing from the insights gained here, new assumptions are posited.

Evidence is Disseminated to Key Interests/Advocates, and the Media, where it is Simplified and Translated

Nutley (2003, p. 12) in discussing research practice refers to the need for research to be translated if it is to get to end users such as politicians: "To have an impact, research findings need to be adapted to, or reconstructed within, practice and policy contexts." She further refers to the need to recruit supportive opinion leaders to move beyond simple dissemination of evidence. Opinion leaders serve a similar role to Moynihan's advocates, people who do the research translation job, interpreting and making sense of the evidence in a way that will attract the attention of politicians.

The media are increasingly playing a similar role to advocates or opinion leaders, with particular consequences. Rist (2007), in discussing the findings of a study into evaluation in the public arena, refers to what happens to evidence as it is disseminated through the media and other channels:

> Organizations such as the Campbell Collaboration that focus on building definitive knowledge assertions from strong empirical foundations are increasingly defining what is or is not factual—and in an epistemological sense—also true. But sharing evaluative information in public life means going in the opposite direction. In the

Table 2
Revised Theory of Change for Politicians' Use of Evidence

Assumption 1	Evidence is produced in an accessible and timely manner
Assumption 2	Evidence is disseminated to key interests/advocates, and the media, where it is simplified and translated
Assumption 3	Evidence is brought to the attention of politicians for consideration, with background analysis done by support staff
Assumption 4	Politicians view the evidence as being of significance
Assumption 5	Politicians use the evidence in discussions and debates surrounding decisions

public arena, the emphasis is on making the message simple and straightforward. It is to be packaged for the message bar on CNN, for a five line summary in *USAToday*, or for one line in a political speech. Complexity and nuance are not the order of the day. One or two "facts" and the bottom line of what works or not (crime prevention programs), what is good or not (comparing test scores in different schools), or the quality of a program (surgical survival rates in different hospitals) are what are shared. What also happens in this reporting within the public arena is that there is a distortion of the data to emphasize simplicity and brevity. Contextual understandings are omitted.

Politicians react to evidence as it is presented in the media. Therefore getting evidence into the media and managing the simplification and distortions that can arise is an important task for those promoting the use of evidence by politicians. There are indications that some players in this field are increasingly aware of this need. The Coalition for Evidence Based Policy in the United States, for example, is a proponent of greater use of randomized trials in evaluating social programs. As part of its effort to influence politicians to make greater use of the findings from randomized trials and to devote more resources to funding randomized trials, attention is paid to simple presentation with compelling specifics, and in doing almost all the work for politicians, such as drafting the language to use in legislation, and giving them specific interventions to implement (Baron, 2006).

Evidence is Brought to the Attention of Politicians for Consideration, with Background Analysis Done by Support Staff

Advocates, opinion leaders and the media all play an important role in bringing evidence to the attention of politicians. But often this attention-bringing is further filtered and tested by officials working for politicians. With regard to ministers, Quinn (2008), as discussed earlier, has highlighted to role of senior officials groups supporting cabinet committees as key evidence channels. More generally, with regard to parliamentary scrutiny of the performance of the executive, the role of support staff for parliamentary committees has been highlighted in this study. Parliamentary committees are often poorly resourced, and their ability to access, test and filter evidence is, accordingly, limited in scope (Johnson and Talbot, 2007). Where staff resources are made available to committees to fulfill their scrutiny role, evaluative evidence on the performance of the executive has more chance of being used, as it is tailored to their particular needs (Nutley, 2003). There is also greater likelihood of evidence being tested and the limitations of poorer quality evidence highlighted.

If everything "right" is done at each step of the theory of change, will evidence be used by politicians? The answer, of course, is that there are no guarantees. And there is no guarantee that evidence will be used in a "rational" manner, with only "sound" evidence used, and actions flowing from the evidence. Evidence will always be only one of the elements politicians consider in democratic decision making. Further, as Sunstein (2005) advocates, ignoring the accepted wisdom as posited by commonly agreed evidence and presenting dissenting views can be an important role for politicians (and others) in producing a healthy democratic society. So a linear relationship leading from evidence to political use of that evidence is not always desired or desirable.

Regarding the overall themes of this book, there is little to suggest that, to date, new forms of evidence such as second order evidence and increased commissioning of evaluations have changed the relationship between politicians and evidence. Sometimes evidence from systematic reviews is accepted and used, other times it is disputed. Those who advocate a hierarchy of evidence with randomized trials at the top, such as the Coalition for Evidence Based policy, still have to work at political engagement if they want their views to be influential. Politicians and evidence can co-exist, but the relationship is not straightforward, and it is one that needs to be continually worked at.

Notes

1. Mayors, aldermen and committee chairs as equivalent to cabinet members, committee members as equivalent to parliamentary committee members, and individual local politicians as equivalent to individual parliamentarians.
2. Deposit Interest Retention Tax (DIRT) was introduced under the 1986 Finance Act as a levy on interest paid by financial institutions to their customers. In 1998 media reports revealed widespread evasion of the tax through the use of fake "non-resident" accounts, with the collusion of several of the state's largest financial institutions.
3. Introduced in 1997 as the Expenditure Review Initiative, the name was changed to the Value for Money and Policy Review in 2006. Under the initiative, government departments must conduct a number of evaluation reviews each year that analyze public spending in a systematic manner.

References

Baron, J. (2006). *Advancing evidence-based reforms in U.S. social policy*, presentation to International Evaluation Research Group (INTEVAL) meeting, May 30, Washington D.C.

Barrington, R. (2007). "Response to *Three weddings and a divorce*." In *National Economic and Social Forum, Evidence-based Policy Making*, Dublin: National Economic and Social Forum: 72-74.

Bogt, H. (2004). "Politicians in search of performance information? Survey research on Dutch aldermen's use of performance information." *Financial Accountability & Management* 20,3:221-252.

Brun, M.E. and J.P. Siegel (2006). "What does appropriate performance reporting for political decision makers require? Empirical evidence from Switzerland." *International Journal of Productivity and Performance Management* 55,6:480-497.

Daft, R.L. and N.B. Macintosh (1978). "A new approach to design and use of management information." *California Management Review* 21,1:82-92.

Department of Health and Children (2002). *Report of the forum on fluoridation.* Dublin: Stationery Office.

Johnson, C. and C. Talbot (2007). "The UK parliament and performance: challenging or challenged?" *International Review of Administrative Sciences* 73,3:113-131.

Likierman, A. (1988). "Information on expenditure for parliament: an overview and future directions." *Parliamentary Affairs* 41,3:362-379.

Lomas, J. (2007). "Three weddings and a divorce—case studies and lessons for the use of evidence in policy making," in *National Economic and Social Forum, Evidence Based Policy Making*, Dublin: National Economic and Social Forum: 59-71.

MacCarthaigh, M. (2005). *Accountability in Irish parliamentary politics.* Dublin: Institute of Public Administration.

Mayne, J. and E. Zapico-Goni (1997). "Effective performance monitoring: a necessary condition for public sector reform." Ch 1 in J. Mayne and E. Zapico-Goni (eds.), *Monitoring Performance in the Public Sector*, New Brunswick, N.J.: Transaction Publishers.

Moynihan, D.P. (2005a). "Why and how do state governments adopt and implement 'managing for results' reforms?" *Journal of Public Administration Research and Theory* 15,2: 219-243.

Moynihan, D.P. (2005b). "Goal-based learning and the future of performance management." *Public Administration Review* 65,2: 203-216.

Nutley, S. (2003). *Increasing research impact: early reflections from the ESRC evidence network.* ESRC UK Centre for Evidence Based Policy and Practice Working Paper 16, Department of Management, University of St. Andrews (http://www,ruru.ac.uk, last accessed 6 July 2007)

OECD (2005). *Modernising Government: The Way Forward.* Paris: OECD Publishing.

OECD (2007). *Towards Better Measurement of Government.* OECD Working Papers on Public Governance, 2007/1, Paris: OECD Publishing

Pollitt, C. (2006). "Performance information for democracy? The missing link?" *Evaluation* 12,1: 38-55.

Quinn, O. (2008). *Advisors or Advocates? The Impact of State Agencies on Social Policy.* Dublin: Institute of Public Administration.

Reville, W. (2007). "Why water fluoridation wouldn't wash if it was being introduced now." *Irish Times*, March 29: 17.

Rist, R.C. (2007). "Bringing evaluation into public life—or bringing public life into evaluation: making sense of this new era," in R. Boyle, P. Dahler Larsen and J. Breul (eds.), *Open to the Public: Evaluation in the Public Arena*, New Brunswick, N.J.: Transaction Publishers.

Sunstein, C.S. (2005). *Why Societies Need Dissent.* Cambridge, MA: Harvard University Press.

Thomas, P. (1985). *The Aims and Outcomes of Social Research.* Croom Helm.

9

Conclusion and Perspectives: Risks and Promises of the Evidence Movement

Tom Ling, Frans Leeuw, and Olaf Rieper

In the introduction chapter we identified a number of inter-related and over-lapping questions to be addressed in this book about the evidence movement. In answering these questions we hope better to understand the nature, origins, immediate impacts and long-term implications of the evidence movement. It should be clear that this is not only relevant to the narrow issue of the evidence movement itself (however defined) but also to much wider issues concerning evaluation, knowledge, and decision-making. These questions were prompted by our experiences as evaluators. These experiences were influenced by the evidence movement in at least two ways. First, there were changing expectations about how research and analysis was used in the conduct of evaluation. Impact Assessments in the European Commission, performance audits in supreme audit institutions, and invitations to tender for evaluations all reflected changing expectations of what constitutes compelling evidence and the acceptable principles of argumentation. Second, the programs and projects we were called upon to evaluate were changing. At best, programs had more clearly identified outcomes and were associated with more specific and measurable impacts. Even without this, program managers and practitioners were more used to collecting evidence of impact and were less surprised to be asked for data about costs, processes and impacts. At the same time, we were unconvinced by all of the claims made on behalf of the new evidence-based world. We were not only concerned that actual practice often failed to meet the high standards set by the evidence movement, but also we were not fully convinced that the evidence movement had fully grasped the density of relationships between evidence, theory, action and benefit. This led us to pose some questions at the start of this book. The questions included the following:

1. Is "evidence-based evaluation" just another term for good quality evaluations, or does it point to a distinctive form of evaluation? If there is nothing distinctive about things labeled "evidence-based," will "evidence" soon lose its semantic power? Can we prevent the word becoming little more than a cheer-leader for a variety of interests or might the term itself become a barrier to further theoretical progress and social betterment?

2. Does evidence in the form of systematic reviews represent a helpful second order of knowledge based on primary order knowledge—i.e., single evaluations and other primary pieces of research? Or is it a double-edged innovation that on the one hand benefits decision-making by focusing it on an extensive evidence-base and on the other hand risks creating arbitrary definitions of what that evidence-base should include and how evidence should be weighted? Do systematic reviews, in the name of objectivity and pragmatism, smuggle in concealed preferences and unfounded epistemological claims?

3. Does the term "evidence-based policy and practice" add anything new? Does this imply that at some time in the past we had "evidence-free" policy and practice? If this idea is absurd then exactly what is claimed to be added by the term "evidence-based" in policy and practice? Is the distinctive thing the focus on outcomes rather than knowledge about process, professional practice and convention?

4. How is evidence used by practitioners, policymakers and citizens in their decision-making? Irrespective of the epistemological and philo-sophical claims made on behalf of the evidence movement, what do people actually do when they claim to be "evidence-based"?

5. There are the methodological issues and controversies behind the evi-dence movement that we need to make explicit if we are to understand the true implications of the movement. When we review these, are we witnessing a new "war of scientific paradigms"? Or, rather, are we observing yet another round of the old methodological debate between the "verstehende" and "erklärende" sciences? What, if anything, is new about the methodological issues and how helpful has the contribution to methodology been?

6. This volume visits a number of policy sectors including health, educa-tion, social affairs, development aid, and criminology. The book also spans many nations and a wide variety of organizations. Does the theory and practice of evidence-based decision making differ systematically across countries, sectors and organizational types?

We conclude this book by considering what the preceding chapters had to say in response to each question.

Question 1: Is "evidence-based evaluation" distinctive, new and beneficial?

This question directs our attention to the world of evaluation. Has the evidence movement introduced something new to the meta-discipline of evaluation? In Chapter 2 Leeuw warns us against adopting an over-simplistic answer to this question by showing the deep historical roots of outcome-oriented and experimental evaluation methodologies. Ling also suggests that claiming the mantle of being "evidence-based" may (amongst other things) be part of a long-running effort by professionals and other experts to shore up their social power. This suggests that we should be cautious about viewing the claim to be "evidence-based" as an unalloyed benefit. And Boyle reminds us that politicians parading their claim to be purely evidence-based might in practice be pursuing other agendas. The meta-discipline of evaluation has therefore found itself caught up in a range of claims and counter-claims.

For evaluation, a significant theme to emerge from this volume is that the evidence movement directly makes claims about (and has changed) what should be evaluated, how it should be evaluated, and how evidence should be weighed in arriving at a conclusion. It therefore provides a basis for judgments about who should be held responsible for what, and how well they have accomplished their tasks. Therefore what counts as evidence, and how it is weighed, will affect not only performance management but also the wider calling to account of those in power. Evidence should therefore be seen as part of the disposition of power and responsibility and changing what counts as evidence not only alters evaluation but also changes the nature of accountability. As we argue below, this has significantly influenced evaluation as a discipline leaving it more fully focused on measurable outcomes and the counter-factual especially through the use of experimental or quasi-experimental approaches. This has resulted in more sophisticated and very possibly more helpful, evaluations. The successful interpretation of evidence by policymakers, practitioners, and citizens can contribute to social betterment and the evidence movement although it is clear that this simple model of change may in practice break down. Even so, reliable evidence of social need and of impacts provides the bedrock for creating and allocating resources for social benefit.

The model of change connecting evidence to social betterment is explored in this volume by Mayne. He considers how results based management in the public sector has opened up the opportunity to have evaluations that systematically examine whether aims have been delivered and to interrogate the claims of public bodies to be doing good. Mayne makes a crucial point in connecting the evidence movement to results management. Results based management brings with it new practices in management. At the planning stage, targets and key measurements will be established; in monitoring data will be collected and analyzed in the light of the expectations set, information will be used to adjust activities and accountability will be largely focused on the attainment of results. This introduces new information and new incentives and evidence is at the very heart of this. On the face of things, this suggests that the confluence of results based management with evidence-based evaluations has a transformative power. This is a power that influences the autonomy of professionals, changes the lives of public servants and introduces new elements into accountability arrangements.

Hansen and Rieper also remind us of the role that second-order evidence can play in supporting evaluations and policy making. They allow evaluators to incorporate existing evidence in a more effective manner and so strengthen the degree of rigor in their evaluations. By locating the specific evidence used to support policy making and evaluation within the context of the wider evidence-base, conclusions, and judgments should be better informed. However, as Hansen and Rieper also show, second order systematic reviews also often seek to set a standard for data collection and analysis which emphasizes the superiority of experimental and quasi-experimental evaluations. Consequently more context specific factors, which involve a particular conjunction of factors, may not be given sufficient weight. This is a consequence of an approach that privileges replicable studies. Launsø and Rieper also remind us that other sorts of evidence—in the form of users experience and expertise—can also be systematically introduced (but this is not always done).

This book resists the idea that there is one simple consequence of the evidence movement for the practice of evaluation. Different disciplines, traditions, professions and political systems have responded to the evidence movement in different ways, sometimes dissipating and sometimes amplifying its impact. Crucially, however evidence is organized (and Ling identifies the examples of dashboards and balanced scorecards) not all behavior will be captured and not all desirable outcomes will be identified. Between the interstices of aggregated evidence there will always be room for gaming, initiative, independent action, professional judgment, bureaucratic discretion, or whatever. Consequently, even where the evidence movement is strong, we may see a gradual softening of the target culture, or at least the view that target culture can best succeed when it also attends to the ideological preferences, interests and powers of practitioners and decision-makers. With this may come a wider view of what we should count as "evidence" and a more modest account of the impact of evidence on progress and social betterment. However, this book is not incompatible with the contention that "evidence-based evaluations" (especially when linked to performance management) have introduced a more "scientific" element to management and accountability and have at least the potential to support improvements in services and in accountability.

Questions 2 and 3: Are systematic reviews and evidence-based policy and practice notable and beneficial innovations?

The promise of systematic reviews is that they represent a form of second order knowledge that has a higher quality than the individual primary studies included in the review. They also claim to generate more systematic and well-judged conclusions than would be arrived simply by reading each of the individual pieces of research. This claim to higher quality is based upon the screening and aggregation of the primary studies. They also have the potential for constructing more developed theories of the mechanisms of the interventions to be studied.

Hansen and Rieper trace the arrival and institutionalization of second-order knowledge producing organizations. They represent a significant innovation and are an important part of the evidence-based movement. They provide a valuable mechanism for managing the burgeoning quantities of evidence available to decision-makers. Whilst some clearly suggest a hierarchy of evidence, others use a matrix approach that shows how different types of evidence might have different strengths and weaknesses.

On balance, however, the usage of the main new product—the "systematic review"—is partial, as Boyle shows. We cannot, therefore, explain the growth in systematic reviews simply in terms of the rising demand for the product. Possibly, the growth has been more producer-led than user-led. Indeed it is clear that many systematic reviews remain underused. As Hansen and Rieper suggest, the rise of the systematic review might be seen at least in part as the result of an alliance of interests amongst researchers, professionals and public bodies. They may support, but apparently rarely drive, improvements in decision-making and in professional practice.

Meanwhile Launsø and Rieper argue that, to the extent that systematic reviews ignore or marginalize user-dependent knowledge, they miss out a crucial element in the evidence-base of good practice. If the rise of the systematic review cannot be wholly explained by the rising demand for it then perhaps it is a "supply driven" product. This might encourage us to consider Ling"s suggestion that the attractiveness of the systematic review is related to the desire to redefine knowledge in a way that re-establishes the power of experts, science and professionals. The systematic review, after all, arrived at a time of growing suspicion of experts, science and professional power. Is it too much to see it as one response to the rise of an anti-science movement and suspicion of experts and professionals?

We have already commented on the absurdity of suggesting a past age when evaluation and policy was supposedly "evidence free" and noted Leeuw's tracing of the historical antecedents of the evidence movement. And we have noted Boyle's cautionary tale of when and how politicians use evidence. However, where the weight of evidence is clear, the systematic review provides practitioners and policymakers with a welcome alternative to having to make sense of conflicting claims and it facilitates the arrival of significant practices such as Technology Assessment, Impact Assessments, professional guidelines, clinical audit (against evidence-based standards), more focused inspections, and more outcome-oriented, "better" regulations. These may not always have been well balanced or perfectly executed but they have clearly added something new (and, we would argue, broadly beneficial) to the quality of decision making.

But we have also identified limitations to the "meta" evidence produced by systematic reviews. We have noted that for evidence to be practicable it often needs to be interpreted and applied in a specific context. Knowing that, for example, that there is systematic aggregated evidence in favor of a particular approach to quality improvement in health care more generally does not guarantee that it will work in a specific context.

Question 4: What are decision-makers really doing when they claim to be "evidence-based"?

Boyle's discussion illustrates that evidence in the form of systematic reviews is not always used by politicians, depending heavily upon the context. A contingency theory of the use of evidence in policy decisions is therefore suggested. The availability of evidence is only one factor influencing decision-making. Mayne suggests that the use of evidence by senior management in outcome-based management is based on a theory of change such that the quality of the evidence is only one factor among several others that determine whether or not the evaluative results are put in use (instrumental or conceptual).

In the chapter by Launsø and Rieper the potential for use of evidence in the form of service users' perception and assessment of the service received is disregarded in the professional context of medicine in which so-called objective findings (that is user-independent findings) are prioritized. This also points to the importance of who is interpreting the evidence— professionals may be more influenced by other professionals than by users' experiences. Vaessen and van den Berg also show that program-level evaluations, which may be useful for accountability purposes, may lack sufficient granularity to support learning for practitioners or decision-makers at lower levels of specific interventions within the program. By aggregating evidence and identifying "universal" lessons evaluations may be unhelpful at disaggregated and local levels. Ling's chapter also supports this conclusion and argues for evaluations to capture local evidence if they are to support such learning. The chapters by both Vaessen and van den Berg and Ling also note how some evaluators have sought to manage this need for understanding both universal mechanisms and local contexts by applying a version of critical realism.

Question 5: How should we understand the methodological claims of the evidence movement?

Leeuw encourages us to take an historical view of the evidence movement. From his chapter it is apparent that not all claims to historical originality should be taken at face value. Indeed, the philosophy of science and epistemology demonstrate a long-running debate about the nature of scientific evidence and its relationship to knowledge. There is a danger that the subtlety and complexity of these earlier arguments could be lost in a wrong-headed binary argument between being "evidence-based" and being not "evidence-based." The particular examples used in this volume demonstrate that the issues in real decision-making setting are concern managing different types of evidence and integrating evidence with values and intuition to arrive at timely and adequate evaluations and decisions. In this sense, the evidence-based movement might be more productively cast as an evidence-informed movement. The former suggests a world where evidence is the arbiter of the most important ("what counts is what works"); while the latter suggests a context where evidence, learning and change enjoy a more messy and iterative set of relationships.

However, the evidence movement already has some enduring consequences. Rieper and Hanson note the innovation of second-order evidence producing organizations, and Mayne recognizes the innovative potential of results based management. Leeuw also identifies the importance of evidence produced through an experimental or quasi-experimental methodology. Together these three elements identify an orientation towards outcomes, the use of experimental methodologies to address the problem of the counter-factual (what would have happened in the absence of the intervention?), and the systematic collection and aggregation of such evidence to provide a significant bank of knowledge to inform evaluations and decisions. This may not constitute a philosophical or epistemological innovation but it does provide a pragmatic and beneficial addition to the tools available to decision makers. The chapter by Launsø and Rieper, for example, indicates a vivid debate on the practical issues in conducting systematic reviews related to the criteria for inclusion and exclusion of primary studies. This debate has evolved with particular energy around the concept of the hierarchy of evidence, mentioned by the chapters of Leeuw and by Rieper and Foss Hansen. The hierarchy of evidence in its various versions represents a ranking of research designs in relation to assessing the effects of specific interventions. However, it is more of a pragmatic response to the explosion of evidence than a well developed philosophical position. But solving this practical problem creates hierarchies of knowledge which allocate greater weight to some sorts of knowledge with inevitable consequences for the allocation of power.

Question 6: How does the evidence movement differ by country, by discipline and by sector?

The chapters in this volume span different sectors such as health, crime and justice and the environment. Accounts have drawn upon cases from countries like Ireland, United Kingdom, Denmark, Canada, and the United States as well as international organizations. This is not a representative sample, but from these illustrations we can see that the effects of the evidence movement vary. In some situations the issue of ranking and weighing evidence for policy-makers is important. Elsewhere policy-makers are more interested in using evidence to conceal their real motives. In some situations professionals are concerned to find ways to integrate evidence into protocols, guidance and audit and so enhance their professional autonomy. Elsewhere, evidence is used to support results based management which may undermine the independence of professionals. In some situations the key issues are methodological and epistemological (for example over inclusion or exclusion in a systematic review) and in other situations the issues are more pragmatic (for example how best to make evidence available in a form that will support decision-makers).

But behind this diversity of effects and issues lies a common core of concerns. The common core is about how to move from a more informal, convention-rich and less transparent approach and create a more formal, methodologically grounded and more transparent approach to using evidence. Where once decision makers formed their judgments based on a mix of experience, intuition, and access to personally preferred evidence, the evidence movement has successfully encouraged a shift towards judgments based upon authenticated, clearly structured, and widely shared sources of evidence.

This, it seems to the authors of this book, is a clear benefit for accountable and good quality decision-making. It supports higher standard of evaluation than would otherwise be routinely possible. Resistance to it is often a reflection of sectional interests or institutional inertia. However, as should be clear from the chapters in this volume, there are also more sophisticated grounds for anxiety. The model of change that suggests that evidence can be collected, agreed, prioritized, and applied as a basis for evaluation and action is naïve. For many significant areas of social and political life, evidence rarely presents itself in such an uncontested manner. Furthermore, applying wider evidence to a particular situation will often require the application of local expertise.

We are very conscious that the chapters in this volume draw upon a variety of wider scholarship. Among other things, the various chapters refer to literature on the sociology of knowledge, new public management, evaluation, the power of professionals and users, and public administration. They also draw upon debates within policy areas such as health, criminology and development. It is this sweep of perspectives that reveals the diverse and complex nature of the evidence movement and its consequences. This book makes specific contributions to the literature identified within the chapters. However, there is also a benefit in stepping back to briefly consider the role played by the evidence movement in shaping such a diverse literature and such varied practices.

The chapters presented here reinforce Davies (2004) assertion that there are many different types of evidence. Davies suggests that these include:

- Impact evidence
- Implementation evidence
- Descriptive analytical evidence
- Attitudinal evidence
- Statistical modeling
- Economic/econometric evidence

- Ethical evidence

Indeed, many others have commented on an apparently growing list of "conceptions of evidence" (Humphries, 2003) and methods for generating evidence (Davies, 1999). Such an eclectic approach raises the problem of how, if at all, to weigh and prioritize evidence (or the methodologies used to generate evidence). Indeed, it is clear from the chapters that evidence sources even wider than those listed by Davies are also drawn upon; including evidence about what stakeholders want, what is electorally acceptable, evidence about expert judgment and so forth. One solution to this diversity of evidence sources is to have a hierarchy of evidence (or methods), but we have also seen that this is challenged even in health sciences, where it enjoys most support. Consequently, there is a tension between the need to avoid not only an eclectic "bricolage" of evidence where each bit of evidence has equal weight, but also to avoid a naïve hierarchy.

Evidence about the costs and benefits of collecting and using evidence is, ironically, limited. Simons (2004) has argued that a reliance on quantitative evidence at the expense of qualitative evidence has damaged professional practice and Sir David King, then the UK Government Chief Scientific Adviser suggested in 2005 "in recent years we have seen the level of public interest in evidence-based issues increase, and in some cases the level of public confidence in government's ability to make sound decisions based on that evidence decrease" (OST, 2005). However, other than evidence of such anxieties from some quarters, there is no systematic evidence base for assessing the costs and benefits of the evidence based movement.

Despite this lack of systematic evidence, it is clear that the evidence based movement has invigorated many aspects of social and political life and energized thinking in important areas. It was previously considered to be a particularly Anglophone tradition (David, 2002), but we have seen a somewhat wider reach in this book to include at least the Netherlands and Scandinavia. Indeed, the evidence based movement and its reception can readily be seen as the latest phase in a longer debate dating back to the European Enlightenment.

In fact it may be better not to see the evidence based movement as a social movement which one is either for or against. Instead, one could see its contribution as being in precipitating a debate about the proper contribution formal and informal evidence to the betterment of society, and to re-open the debate about how to prioritize and weigh evidence in processes of argumentation, decision-making and judgments. In particular, it has focused attention onto the systematic review as an additional source of evidence. If the aim of the movement was to build a consensus in favor of one set of rules governing epistemology and methodology then it has failed. If, however, the aim was to provoke a mature debate about the role of evidence in improving society then, in our view, it has succeeded. We hope that this book makes its own contribution to this debate.

References

David, M.E. (2002). "Themed section on evidence-based policy as a concept for modernising governance and social science research," *Social Policy and Society* 1,3: 213-214.

Davies, P. (1999). "What is evidence-based education?" *British Journal of Educational Studies* 47,2: 108-21.

Davies, P (2004). *Is Evidence-Based Government Possible?* Jerry Lee Lecture, 4th Annual Campbell Collaboration Colloquium, Washington DC.

Humphries, B. (2003). "What *else* counts as evidence in evidence-based social work?" *Social Work Education* 22,1: 81-91.

OST (Office of Science and Technology) (2005). *Guidelines on scientific analysis in policy making: a consultation by the government chief scientific adviser.* London: OST.

Simons, H. (2004). "Utilizing Evaluation Evidence to Enhance Professional Practice." *Evaluation* 10,4: 410-429, London: Sage.

Contributors

Richard Boyle is a Senior Research Officer with the Institute of Public Administration in Ireland. He has worked with the IPA since 1986, having previously worked on policy and evaluation issues in local government in England. His research interests focus on public service modernization, managing for results in the public sector, and developing and implementing effective performance management and evaluation systems. He has researched and written extensively on public service reform, performance measurement and evaluation. He has worked widely with Irish central and local government, the OECD, and the World Bank. He was a member of the Board of the European Evaluation Society from 2002 to 2005 and is chair of the Irish Evaluation Network.

Hanne Foss Hansen is professor in public administration and organization at Department of Political Science, University of Copenhagen. She has a Ph.D. degree from Copenhagen Business School 1986 on organization and social control in academia. Since 1993 she has worked at the University of Copenhagen. She has published books and articles on organizational effectiveness in public organizations, public sector reforms, reforms in higher education, regulation, evaluation models, and evaluation practice. She has also worked with evaluation in practice especially with auditing and accreditation in higher education. She is currently involved in a research project (together with Olaf Rieper, The Danish Institute of Governmental Research, AKF) on meta-evaluation and the idea, institutionalization, and methodological practices of the evidence movement.

Laila Launsø's field of research has since 1976 been concentrated on conventional and alternative treatments based on both users' and practitioners' perspectives, and on bridge building between conventional and alternative practitioners. She is director of a research-based evaluation of a team-based treatment for people with multiple sclerosis initiated by the Danish Multiple Sclerosis Society and senior researcher and project director at the National Research Center in Complementary and Alternative Medicine, The University of Tromsø, Norway. She has a doctoral-degree in Sociology 1996 and was visiting professor at the University of Michigan in 1988 and connected to the pharmaceutical faculty at the University Claude Bernard in Lyon, France in 1995/1996.

Frans L. Leeuw is the Director of the Justice Research, Statistics and Information Center, affiliated to the Dutch Ministry of Justice and Professor of Law, Public Policy and Social Science Research at Maastricht University, the Netherlands. He is a sociologist (Ph.D, Leyden University, 1983). Earlier he was Professor of Evaluation Studies at Utrecht University, Director of the Performance Auditing and Evaluation Department of the Dutch National Audit Office, Dean of the Humanities and Social Sciences Faculty of the Netherlands Open University, Chief Inspector for Higher Education in the Netherlands and associate professor of policy studies at Leyden University. He has been one of the founding persons of the European Evaluation Society EES and past president of this organization. Currently he is President of the Dutch Evaluation Society. He is a Faculty member of the IPDET program. He has worked for the World Bank, the EU and for many agencies and ministries of the Dutch government.

Tom Ling is Director for Evaluation and Audit at RAND Europe. He studied Social and Political Sciences at Cambridge University and completed a PhD in Government at Essex University. He joined RAND following four years as Senior Research Fellow at the National Audit Office in the UK. He has worked on evaluation projects with the European Commission, UK Government Departments, the National Audit Office, The Health Foundation and many others. He has published widely on evaluation, accountability and related topics. He is currently working on a Public Audit Handbook, a multi-site program evaluation of quality improvement in healthcare, and an analysis of the Impact Assessment process in the European Commission. He is Professor of Public Policy (Emeritus) at Anglia Ruskin University where he contributes to postgraduate teaching.

John Mayne is an independent advisor on public sector performance. He has been working with a number of organizations and jurisdictions, including the Scottish Government, the United Nations, the International

Development Research Centre, the OECD, the Asian Development Bank, the European Union, and several Canadian federal organizations on results management, evaluation and accountability issues. Until 2004, he was at the Office of the Auditor General where he led efforts at developing practices for effective managing for results and performance reporting in the government of Canada, as well as leading the Office's audit efforts in accountability and governance. Prior to 1995, Dr Mayne was with the Treasury Board Secretariat and Office of the Comptroller General. He has authored numerous articles and reports, and edited five books in the areas of program evaluation, public administration and performance monitoring. In 1989 and in 1995, he was awarded the Canadian Evaluation Society *Award for Contribution to Evaluation in Canada*. In 2006, he became a Canadian Evaluation Society Fellow.

Olaf Rieper has an M.A. in sociology and a Ph.D. from the Copenhagen Business School. He is Director of Research at the Danish Institute of Governmental Research (AKF). He has conducted evaluations and research on public programs for Danish and Norwegian government agencies and for the EU Commission. He has been visiting scholar at the School of Social Work, University of Michigan, and has worked at the Centre for European Evaluation Expertice (C3E) in Lyon. He has written books and guidelines on evaluation methodology and research methods and was the first elected chairman of the Danish Evaluation Society.

Jos Vaessen is a researcher at the Institute of Development Policy and Management at the University of Antwerp. He studied Agrarian Development Economics at Wageningen University, obtaining his Master Degree in 1997. Over the last 10 years he has worked in research and teaching within the broad domain of development intervention. During this period he has done extensive field work in Central America (mainly in Nicaragua, Guatemala and Costa Rica) on topics such as rural livelihoods, rural finance, sustainable agriculture and natural resource management. He has also participated in several policy research and policy evaluation activities on these and other issues, including an ongoing collaboration with the Evaluation Office of the Global Environment Facility as an external consultant. He regularly publishes on the abovementioned issues as well as evaluation methodology, in particular the application of theory-based evaluation in development interventions.

Rob van den Berg studied contemporary history at the University of Groningen in the Netherlands. Since 2004 he has been the Director of the Evaluation Office of the Global Environment Facility. Prior to this position, he worked for the Dutch Ministry of Foreign Affairs for 24 years in various positions within development cooperation and other policy domains. From 1999 until 2004 he was Director of the Policy and Operations Evaluation Department of the Dutch Ministry of Foreign Affairs. From 2002 to 2004 he was the chairman of the OECD/DAC Network on Development Evaluation. He chaired the Steering Committee of the joint international evaluation of external support to basic education, which published its final report in 2003. He has also served as the Executive Secretary of the Netherlands' National Advisory Council for Development Co-operation, and as the Head of the special program for research of Dutch development cooperation. He is a member of faculty of the International Program for Development Evaluation Training (IP-DET) since 2000.

Index

For Product Safety Concerns and Information please contact our EU
representative GPSR@taylorandfrancis.com
Taylor & Francis Verlag GmbH, Kaufingerstraße 24, 80331 München, Germany